Ending Fossil Fuels

Holly Jean Buck is Assistant Professor of Environment and Sustainability at the University at Buffalo, New York. Her book *After Geoengineering: Climate Tragedy, Repair and Restoration* explores best and worst-case scenarios for climate intervention.

Ending Fossil Fuels

Why Net Zero Is Not Enough

Holly Jean Buck

VERSO
London • New York

First published by Verso 2021

1 3 5 7 9 10 8 6 4 2

Verso
UK: 6 Meard Street, London W1F 0EG
US: 20 Jay Street, Suite 1010, Brooklyn, NY 11201
versobooks.com

Verso is the imprint of New Left Books

ISBN-13: 978-1-83976-234-5
ISBN-13: 978-1-83976-236-9 (UK EBK)
ISBN-13: 978-1-83976-237-6 (US EBK)

British Library Cataloguing in Publication Data
A catalogue record for this book is available from the British Library

Library of Congress Cataloging-in-Publication Data
Library of Congress Control Number: 2021941206

Typeset in Sabon by MJ & N Gavan, Truro, Cornwall
Printed and bound by CPI Group (UK) Ltd, Croydon CR0 4YY

Contents

Introduction

Controlled Demolition

Explosives toppled the three concrete smokestacks of the Navajo Generating Station. Against the red mesas outside Page, Arizona, the 775-foot towers had loomed, high as 77-story buildings. Technicians had already cut the stacks' backsides, structurally weakening them by removing small blocks.

In the widely circulated drone footage, the towers slowly drift down, in sequence: all three aligning, like some kind of demolition ballet in the desert morning, leaving roiling, sunrise-tinged clouds of gray dust.

For forty years, the 2,250-megawatt plant sited on Navajo land sent electricity to Los Angeles and powered pumps that sent water to Phoenix. It was an engine fueling the growth of the American West, while many Navajo households lacked electricity.

After a few years and $150 million of restoration funds, this thousand acres of Navajo land will be returned to something like a natural state, if all goes to plan. Over 100,000 tons of steel, along with volumes of copper and aluminum, will be salvaged. The stacks will remain in the landscape, ground to pieces and used to rebuild contours that mimic nature.

Video clips of the demolition were shared on social media and with observers around the world. The symbolism said: Coal is done. The demolition is just a few seconds in a larger story. As Nicole Horseherder, executive director of the grass-roots organization Tó Nizhóní Ání, said in a statement on the demolition, "That chapter is now closed, but the work is far from over." The coal mine that fed the plant needs to be cleaned up. "We have to secure water and electricity for many communities that lack access to both. We have to replace the millions of dollars in lost coal revenue from the abrupt closure of the plant and coal mine. And we have to make sure investment flows back into building a more sustainable economy for the Navajo and Hopi."[1] The fall of the towers was just one brief moment in broader struggles: to create a just transition for workers; to rectify the government's relationship with Indigenous communities, and to make sure new energy sources don't fall into the same extractive patterns.

The plant also emitted hundreds of millions of tons of carbon dioxide during its lifetime. Roughly a quarter went into plants and trees, a quarter of it is acidifying the ocean, and half of the carbon dioxide is still hanging in the atmosphere, creating climate chaos. The plant's physical form may be reduced to rubble, but it has a second life as a silent, invisible actor in the atmosphere, reaching around the globe.

Yet the toppling of those smokestacks did represent a visible moment in a wave of coal retirements across the United States —and the world. "Around the world the mood is shifting," began a 2020 cover story in *The Economist,* entitled "Time to Make Coal History."[2] Coal consumption has been slowly declining for years. Companies like Samsung and General Electric have announced an end to building coal plants; funders like BlackRock and the Asian Infrastructure Investment Bank have declared they will stop financing them. From Ontario to South Wales, from Chile to Spain to Beijing, the exit from coal has begun. Britain shuttered a third of its remaining coal-fired generating capacity in the first half of 2020; the last coal-fired power plants there will close by 2024.

Admittedly, it's an uneven shift. On one hand, coal consumption in Asia has grown in the past decade, and it now accounts for 77 percent of all coal use; 2/3 of this is burned in China.[3] On the other hand, Xi Jinping has adopted a net-zero by 2060 target for China. The Philippines declared a moratorium on new plants; Japan, Bangladesh, and Pakistan are all moving away from coal. The UN Secretary-General, Antonio Guterres, called for coal-fired electricity to be eliminated worldwide by 2040 and for rich countries to phase it out by 2030. There is a global consensus that coal is ending. Climate advocates are hopeful that it can be a gateway drug to the phaseout of oil and then gas. Even high-level officials, like the President of the European Investment Bank, are now ready to make declarations like, "To put it mildly, gas is over."[4]

Cheap natural gas, driven by a technological revolution in getting gas out of shale rock, has been a key driver in coal's demise. Increases in solar and wind have also played a role. But climate action movements have also been instrumental in moving this reality forward, *delegitimizing* fossil fuels. Resistance and divestment campaigns led by Indigenous activists and young people have helped create the sense of fossil fuels as an immoral and dangerous investment. Environmental justice communities living near fossil fuel facilities have fought to close down plants and resist new ones.

Fossil fuel company executives refer to climate action protests with an implicit sense of panic. There's a growing sense that the jig is up—or at least, that demand for fossil fuels is peaking, and that includes oil and gas. Reserve-based lending (long-term lending collateralized with the oil and gas reserves underground) to exploration and production companies for new oil and gas developments is down 90 percent from its peak; national oil companies are also moving away from international expansion.[5] Shareholder activism, in terms of pushing management toward incorporating climate change into business plans, is at least partly to blame. Larry Fink, the head of BlackRock—the world's largest investment company, with $7.5 trillion under management—shook the financial world when he stated that "climate change has become a defining factor in companies' long-term prospects" and that BlackRock would require companies to disclose climate-related risks.[6]

The response to the coronavirus also forced a rethink, catalyzing a structural fall in global coal demand, according to the International Energy Agency.[7] Coal demand will not return to precrisis levels if the policies we currently have continue, though the agency cautions that it is too soon to see a rapid decline in oil without a significant shift in policies. Even gas receives a wary eye—the European Green Deal talks about "the phasing out of fossil fuels, in particular those that are most polluting" and also "the rapid phasing out of coal and decarbonizing gas" (which is notably not the same as *eliminating* gas).[8] A key political struggle in the 2020s will be extending the phaseout of fossil fuels to gas.

Taken together, all of this seems to say: We are on the right track. Enough powerful institutions are falling in line, repeating the message, sending the right signals, creating a version of the future that feels both coherent and realistic. We see the peak; we may be passing over the peak. If climate change is a drama, this could be the climax; the rest of the work will feel like resolution.

So is the road mapped, the work pretty much underway? Yes—and no. The dominant thinking still sees the energy transition as happening naturally, with a guiding hand from government. Superior alternatives will come online and be cheaper than fossil fuels; for some uses, they already are cheaper. It sounds great, and it might be the case for the power sector, but the timing doesn't match up for all the pieces. To meet targets, fossil fuel plants will have to be retired before the end of their life; internal combustion engines in automobiles will have to be taken off the road; gas stoves will have to be removed from houses; industrial facilities will have to be retrofitted.

California will ban internal combustion engines by 2035. So will Japan. Officially stated goals, declared with much fanfare, point toward those retirements and policies. Still, concrete plans are often either missing or at odds with the goals. The "production gap" between what countries and companies are planning to produce, and what's needed to maintain a 1.5°C pathway, is stark, per a 2020 report.[9] Currently, fossil fuel companies are still investing trillions. To limit warming to 1.5°C, countries would need to decrease fossil fuel production by 6 percent a year over this decade. But they are planning to increase production 2 percent a year. A 1.5°C pathway would see production consistent with a mere 15 gigaton (Gt) of CO_2 by 2040. But current plans and projections would lead to an equivalent of 40 Gt CO_2 per year (above current levels of about 33 Gt, a figure that doesn't include other greenhouse gas emissions). Depressingly, according to the report, as of November 2020, nations have committed $233 billion in Covid-19 relief funds to fossil fuels, compared with $146 billion to renewables.

What are we to make of this discrepancy between the sensibility that the end is coming and the reality of planned production? The bottom-up delegitimization of fossil fuels as a social norm—the spreading sense that they are dirty and deadly—can only go so far. People live in completely different social realities, polarized by algorithms and inhabiting different filter bubbles. This separation means that the social and cultural transition that needs to accompany the technological transition can't happen "naturally" at all. The social transition is constrained by the platforms that now mediate and determine social discourse. Twitter, Facebook, YouTube: their algorithms aren't built for nuanced, dialogic content that will allow people to gradually change their minds and question fossil fuels. They will show people content that is more extreme, with more emotional cues, designed to appeal to what a person has already agreed with. The political economy of the platforms depends on dividing the public in half and enraging one side with the other side's views, which increases time-on-site and supports the ad-driven business model of the platforms. This engineering of the media and information landscape puts hard constraints on the bottom-up delegitimization of fossil fuels. At best, half of the people can be convinced that they are dangerous and dirty, while the other half will become more entrenched in their position that the other side is using this narrative for political ends.

This moment has a special peril. There's a general danger—that of mistaking the switch in discourse, in the consensus of thought, for a shift in material reality. Of thinking "it feels like things are changing, that everyone is thinking the same thing, so they must be changing." But different versions of reality can coexist side by side. Consider driving through the United States during the coronavirus pandemic. In some places one distinct sense of reality applies. Everyone is masked, they stay away from you in the grocery line with an intuitive sense of what six feet is, businesses are shuttered, there's a sense of existential gravity and grim responsibility in the air. Across a state line, or sometimes just a few miles into the countryside, another sense of reality applies, like crossing a force field. In this reality, there's no major danger beyond a seasonal respiratory virus.

This intimate coexistence of realities that are differently mediated is part of why Russian oil company Rosneft can make statements about "modern environment-friendly fuels and minimization of its environmental footprint" or "preserving the environment for future generations" and offer a "strategic carbon-management framework."[10] Companies will say what they need to, because fundamentally they know the Sustainable Development Goals word salad does not matter; they are making a cynical bet that their reality of continued reliance on fossil fuels will win out. The collective delusion of greening and the shared social norms around it may include a large number of people—but it still has little bearing on material reality.

Mistaking signals and discourse for changes in material reality has grave effects on whether fossil fuels are or are not on their way out. There is a fantasy that elites with PhDs are immune to creating shared stories about what is going on in the world; that they dwell in the world of "science" compared to the masses who are subject to filter-bubble groupthink. But climate policy experts may be perfectly capable of dreaming up a shared storyline about the way the future will go. Acting as something of a check on this are activist groups who are very attuned to mistaking signaling for reality, who for decades have been pointing to greenwashing practices and the "false solutions" of carbon offsets. Policy scholars are also pointing to the divergence between signaling and material changes—Danny Cullenward and David Victor call out the "Potemkin villages" of carbon markets in their recent book, *Making Climate Policy Work*, which is a call to the climate policy elite to step back from groupthink about climate markets.[11] More and more people will be observing and quantifying the gap between discourse and action.

These concerns about the media ecology, cultural politics, and knowledge economy in which we live and breathe are all ambient concerns. There's no simple way of dealing with them. What's clear is that the bottom-up delegitimization of fossil fuels has to be complemented by the top-down planning of phaseout, with broad goals translated into a series of concrete steps. Actually ending fossil fuels is a delicate planning task. It will need regulation and top-down control, not just "natural" market forces.

Get Ready to Ask and Answer Hard Questions

How long should we use fossil fuels? How will we end their use?

In the 2020s, these may sound like easy and simple questions. A carbon budget for limiting global average temperature rise to 2°C requires most global fossil fuel reserves to remain in the ground; that means a third of oil, half of gas, and over 80 percent of coal.[12] And we know that 2°C is already a dangerous level of warming. It's beyond obvious: we should stop using fossil fuels right now, and we need to put fossil fuel companies out of existence. "Leave it in the ground." *Next.*

Yet these questions are not easy or simple. Three realities complicate leaving it in the ground.

Hard reality #1: Despite the rapid growth of renewables, fossil fuels still provide more than 84 percent of primary energy. Renewables (wind, solar, and biofuels) are growing fast but still provide just 5 percent, with hydro providing 6.4 percent and nuclear 4.3 percent.[13] At the same time, about 770 million people in the world have no access to electricity, and 2.6 billion don't have access to energy for clean cooking.[14] The challenge is to increase access to energy while also decarbonizing it. Ceasing fossil fuel use before there are other options to replace this 84 percent of global energy use brings real risks of continuing or worsening energy poverty. This is often recognized by climate advocates—for example, when 432 environmental groups from 53 countries sent a letter to officials in the Biden Administration calling for an end to US financing for the entire fossil fuel supply chain, they also noted that "very rare exceptions for fossil fuel projects intended solely for domestic energy consumption only in Least Developed Countries could be considered, but only after a thorough scenario analysis of all viable alternatives for meeting energy access demonstrates clear necessity and no viable alternatives."[15] It will be necessary to maintain this nuance and global perspective when discussing how to end fossil fuels.

Hard reality #2: Over half of global oil production, and even more than that of reserves, is owned by national oil companies. These are fully or majority-owned by governments.[16] We may think of the household brands we see at filling stations—Shell, BP, Exxon, Chevron, Total, and others—but most oil companies are national oil companies, like Petrobras, Sinopec, National Iranian Oil Company, and Pemex. Shouldn't this make it easy to end production? If the people own these resources, can't we just stop using them? But it also makes it hard, because governments derive income from fossil fuels in myriad forms, from licenses to taxes on production and consumption. Turning away from those revenues in an unplanned fashion, as Nigeria had to do when the response to coronavirus cut production by 25 percent, puts government-funded services at risk.

Hard reality #3: Millions of people work in the fossil fuel industry. In the United States in 2019, oil and gas employed nearly 603,000 and 271,000 respectively, with coal jobs totaling about 74,800.[17] (For comparison, solar employs about 240,000 and wind 120,000.)[18] These jobs are often well-paying and support other jobs in the broader community.

Given all that, how do we end fossil fuel production to meet climate goals in the next few decades? The answer from policymakers seems to be: we don't end it completely. Right now, the dominant policy framework and way of thinking about the challenge, which is net-zero emissions, lends itself to the scenario of continued and significant fossil fuel use. Net zero implies balancing some amount of remaining greenhouse gas emissions with some amount of carbon dioxide removals. In the past couple of years, net zero talk has sprouted up everywhere in corporate sustainability discourse as well as national goals.

Those of us who want a complete end to fossil fuels have to take seriously and prepare thoughtful answers to a difficult set of questions. If producing lower carbon fossil fuels can preserve livelihoods for families and stability for communities, isn't that better than going completely to zero? Wouldn't some amount of continued fossil fuel production, such as for gas peaker plants, avoid the risks of supply disruptions from an unbalanced grid? Or consider this: a recent analysis of how Los Angeles could be 100 percent renewable found that the costs for getting to 80 to 90 percent renewable energy are roughly the same for different scenarios, but they diverge for the last 10 to 20 percent depending on whether peaker plants are run on biofuels and a limited amount of gas with renewable energy credits, versus green hydrogen for the fully renewable scenario.[19] The study also found nearly no additional air quality or public health benefits for making that last 10 to 20 percent of energy demand renewable. Could those extra tens of billions for the 100 percent renewable system be better deployed to solve other environmental or social challenges? A community might well prefer more money for schools or other infrastructure plus a 90 percent renewable energy system over a more expensive 100 percent one. These are the questions and tradeoffs that will have to be debated at local and regional levels.

Then there are the global questions: The United States or Canada or the United Kingdom could phase out fossil fuels, but what kind of difference to the global climate would that make if other countries are unable or unwilling to? How would phasing out fossil fuels in the global North affect the global South? Would production just go there? Isn't there a risk that ending fossil fuels too soon, while billions still don't have access to energy, will keep people in poverty? Is the dream of being fossil free just for environmentalists in rich countries?

Many people are thinking about how to answer these questions. Climate advocates, policymakers, scientists, and others are vigorously talking about the people and communities impacted by the shift away from fossil fuels. The "just transition" conversation has become mainstream.

But there's also a pressing need for people in climate movements to continue to detail plans for phaseout and mainstream them, because the class of climate professionals is still too slow to do it. Climate justice and environmental justice movements have been shouldering this burden for years. Calls to keep it in the ground have resonated since anti-extraction movements in the mid-1990s in Nigeria and Ecuador. Groups like Oilwatch and Greenpeace took up these calls, and eventually the Leave It in the Ground Coalition emerged, which then went on to campaign against fracking in the 2010s.[20] In 2016, the Lofoten Declaration for a Managed Decline of Fossil Fuel Production was created when groups came together in Lofoten, Norway, to advocate for managed decline. The movement has focused some attention to policies targeting the supply of fossil fuels, from national bans to treaties to manage decline. At the same time, environmental justice advocates have focused attention on closing fossil fuel plants and infrastructure, and even when often underresourced, they have produced plans for doing so. Consider, for example, *The Fossil Fuel End Game: A Frontline Vision to Retire New York City's Peaker Plants by 2030*, an analysis by a coalition of environmental justice groups that spells out a roadmap for phasing out gas peaker plants in the city and prioritizes retirements in environmental justice communities.[21] Yet at the same time, while "supply-side" climate policy is becoming more common, it's still entering the mainstream at a glacial pace relative to how acute the need is. Professionals have gotten comfortable with talking about early retirement of coal infrastructure, and we may be at an inflection point where talking about the managed decline of all fossil fuels is a topic of polite conversation at "high-level" banquets. But there remains the challenge of following through with the talk.

Ending fossil fuels demands navigating the risk of energy poverty if replacements don't come online fast enough, the dependence of governments on fossil fuel revenues, and the dependence of communities and households on fossil fuel jobs. On top of these fundamental challenges, there are also two other unsettling scenarios for this decade that may complicate any planned phaseout. One scenario is a coming disenchantment with renewables. The other is the arrival of low-carbon fossil fuels. Both of these prospects will make it extremely hard to completely end fossil fuels. Instead of a peak and steep decline to zero, it's likely we'll see a peak and a sloping plateau where fossil fuels remain a significant percentage of our energy mix.

Prospect One: Renewable Disenchantment?

Renewables have had a meteoric rise and generally positive vibes. It may seem pessimistic to talk about a coming disenchantment. The issue is simply that as renewables scale up, some of their limitations become apparent—issues around land use, materials, and intermittency that were easier to gloss over when they provided a smaller portion of the energy mix. Powering our world with renewables involves transforming whole landscapes for energy production. I'm in full support of this, but I've also spent enough time talking with people in rural landscapes that I can understand the ambivalence or opposition that some people have.

We're starting to be able to envision what a full-on effort toward phasing out fossil fuels and replacing them with renewables might entail. Let's take the example of the United States, which despite being a laggard on climate is actually fairly well-poised to decarbonize. After two years of work, a research group at Princeton put together a monumental study called *Net-Zero America: Potential Pathways, Infrastructure, and Impacts.*[22] It's monumental because it is spatially explicit, and it looks at various dimensions of the pathways to net zero (health impacts, employment, and so on). The study features five different scenarios for getting to net zero by 2050 in the United States, juxtaposed with a business-as-usual reference case. It spells out what needs to be done. All of the scenarios feature coal disappearing by 2030. Seven hundred coal mines close, and 500 coal-fired power plants are retired. Oil and gas decrease between 65 percent and 100 percent. Half a million gas wells close in the 2020s, with $25 billion for cleanup costs.

In this future United States with no domestic fossil fuel consumption, the rollout of wind and solar that supports the electrification of everything requires tremendous land and resources. The rates for deployment would set new records each year; by 2050, there would be 5.7 terrawatts (TW) of wind and solar capacity, at $6.2 trillion of investment. That's just a bunch of numbers, but here's something to pay attention to. In this scenario, wind and solar farms span an area of 1 million km^2. That's equivalent to the combined land areas of Arkansas, Iowa, Kansas, Missouri, Nebraska, and Oklahoma (or an area roughly the size of France and Spain). This is all land that's visually impacted by wind turbines, not where the turbine pad directly sits, but it's still a stunning amount of land. Land for solar is equivalent to the size of West Virginia (an area about half the size of England), and the entire Atlantic Coast is lined with offshore wind farms.

What kind of politics is necessary to see that happen? On one hand, one could argue that much of the United States is optimized for agricultural production, and it would simply be a matter of adding a wind turbine layer on top of corn and soy fields. Two stacked layers of extraction for export. But the concerns people have aren't simply about the "character of the land"; they also have quite reasonable concerns about the materials used in solar panels and wind turbines and effects upon wildlife. Questions about local benefit and local agency run deep in these communities.

Turning vast portions of the United States into renewable energy production zones will also require a lot of materials. This has been the subject of recent scientific work attempting to quantify the material use, as well as critical work by scholars like Thea Riofrancos, whose research examines the extractive frontiers of lithium; Dustin Mulvaney, who examines the ecological and social dimensions of solar power production; and Ingrid Behrsin, who examines how claims about renewability are produced, to name a few.

It's not just scholars and scientists making observations about the sustainability of electrifying everything. Sitting at a bar in Dodge City, Kansas, I got to talking with a guy who found wind energy to be an unsustainable joke. Over a locally brewed draft beer, he explained that he had been an engineer on the railroad, hauling wind turbine blades from Amarillo, Texas, to some offloading point in Oklahoma. The blades had been manufactured in Vietnam and had a long journey by sea and rail before reaching whatever field in Kansas they would eventually be stationed in. "How are they going to be low-carbon if they come all the way from Vietnam?" he asked. I had no idea what the life cycle analysis would be, but I looked into it later. Many wind turbines are shipped from overseas—the port of Corpus Christi in Texas had to lay down extra track and space to accommodate wind turbine parts. Transport and installation is actually a moderate part of the life cycle emissions of a turbine, along with manufacturing. One recent life cycle analysis of emissions from wind power found around 5 grams of CO_2-equivalent per kilowatt hour of generation, which is on the low end of the range of estimates.[23] Many analyses report a carbon payback time of 7–9 months; that is, if a turbine runs for longer than that (say, its expected twenty years), the clean energy it's providing far outweighs the emissions. Wind energy really is a winner, all things considered. But the point here is that (a) the emissions are more than zero and (b) those numbers are really hard to track down. It's no wonder, then, that the former railroad worker in Dodge City would have wondered about the sustainability of the enterprise, because there's no transparent, easy-to-find analysis of all this. Nobody bothered to brief him.

Because the promotion of solar and wind has largely been technocratic and driven by people in cities, it has missed some of the social challenges in building that much infrastructure. It has also missed many of the opportunities to communicate with different publics about this infrastructure and to learn from them. So there's a risk that community-supported rejection of this infrastructure will limit its needed scale. One recent report out of Columbia University's law school cataloged 100 ordinances in thirty-one states to block or restrict new wind, solar, and other renewable energy facilities—and that's with only a fraction of the needed infrastructure proposed to date.[24]

A worst-case disenchantment scenario a decade or so out is this: the system has leaned heavily into renewables, but it has not managed the necessary planning well and can't provide enough energy. Outages or limited service could lead to tremendous backlash against efforts to decarbonize. Blackouts in California during peak use in heat waves are already producing grumbles (rightly or wrongly, as natural gas power plants can also face difficulties operating in high heat). Power outages during a freezing spell in Texas in February 2021 were blamed on wind turbines freezing up, when in fact most of the capacity that went offline was from thermal power plants that weren't able to run in cold temperatures because they hadn't been winterized.[25]

These dynamics are specific to the United States, but may also apply in other countries. The fact remains that the whole conversation about our future energy system in the global North has taken place in the context of always having enough energy. Mismanagement of the transition risks crushing disappointment about renewables and possibly demands for more "reliable" fossil energy again. A spreading sense of disenchantment about renewables later in the 2020s may make the dream of ending fossil fuels seem naive.

Prospect Two: Clean Fossil Fuels?

While climate activists have been campaigning to leave carbon underground, fossil fuel interests and policymakers have created greener, recyclable fossil fuels. Decarbonization of fossil fuels sounds like an oxymoron, like doublespeak. But it's receiving increasing attention. You may have been hearing more about "low-carbon energy," because it can include fossil fuels. "Low-carbon" has replaced carbon-free or zero-carbon as an aspiration.

"Carbon management" to produce low-carbon fossil fuels is a means of producing fossil fuels with lower carbon intensity (lower associated emissions per unit produced). Decarbonizing fossil fuels includes detecting methane leaks and capturing methane, decarbonizing the transport of the fuels, electrification and carbon capture and storage in refining, and more; there are opportunities to decarbonize across the process. With carbon capture and storage, zero-emissions fuels and zero-emissions fossil-fueled power are both possible. For example, the new Allam Cycle process uses a new kind of turbine that runs on pressurized carbon dioxide, and the process mixes natural gas in oxygen rather than air, with the byproducts being water and CO_2 which can be sequestered underground. These plants do not produce nitrogen emissions, either. Some companies also have plans to pair direct air capture (machines that suck carbon from the ambient air) with enhanced oil recovery (injecting CO_2 into depleted oil wells to get more oil), which could theoretically produce carbon-negative oil. These new technologies and approaches, which range from pilot-stage to proven, mean that lower carbon fossil fuels are becoming a reality.

Policy is already enabling such fuels: the mechanism for turning discourse into material reality is grinding into motion. In the United States, omnibus legislation in December 2020 to fund the federal budget and provide coronavirus relief was passed. Wrapped up in it was a fairly significant Energy Act. Its titles on "Carbon Management" and "Carbon Removal" were aimed at lowering emissions from fossil fuel production as well as "developing carbon removal and utilization technologies, products, and methods that result in net reductions in greenhouse gas emissions, including direct air capture and storage, and carbon use and reuse for commercial application."[26] By the time this book is published, this work will be well underway. And it's not just the United States: Aramco has embraced a "circular carbon economy" framework, which is a mashup of carbon reduction, carbon capture and re-use (including enhanced oil recovery), carbon recycling into synthetic fuels, and carbon removal.[27] The European Commission's Circular Economy Action Plan talks about incentivizing "the uptake of carbon removal and increased circularity of carbon." Countries might not be exactly on the same page with all of these terms. But they may lead toward treating carbon as something to be cleaned up and re-used, rather than eliminated.

Companies, too, are starting to conceptualize oil and gas as differentiated products, meaning that low-carbon oil or low-carbon gas is its own special kind of certified and branded fossil fuel. Low carbon fuel standards, like those in place in California, offer policy support for all this. For example, in January 2021, Occidental Petroleum issued a press release that it had just delivered the world's first shipment of carbon-neutral oil to Reliance Industries in India—the first petroleum shipment where the emissions were offset. This was just using offsets under a certified standard, but it was portrayed as a first step in creating a market for "climate-differentiated crude" as well as a step toward net-zero oil.[28] This shipment was testing the waters, so to speak.

It is worth mentioning hydrogen here, as it can either be an enabler of a fully renewable energy system, or a means of decarbonizing fossil fuels if blended with natural gas. Either way, the climate politics of the next decade will involve a lot of discussions about hydrogen. Hydrogen is hard to wrap one's head around because there are so many ways of producing it and so many use cases. Right now, hydrogen is produced from natural gas in a process called steam methane reforming, which is called "grey hydrogen." if this is done with carbon capture and storage, it is "blue hydrogen." Blue hydrogen is often imagined as a stopgap on the way to developing more affordable "green" hydrogen, which uses wind and solar energy to split water molecules into hydrogen and oxygen. Hydrogen can be stored as a gas or liquid and kept in large salt caverns for months. Then it can be used for a variety of things—for transportation, in industrial processes, or in conversions back to power ("power-to-gas-to-power; P2G2P"). Because hydrogen has these multiple use cases and can be stored, proponents in the Hydrogen Council suggest that "the years 2020 to 2030 will be for hydrogen what the 1990s were for solar and wind."[29] Basically, full decarbonization roadmaps without nuclear or carbon capture and sequestration tend to lean on green hydrogen for things like decarbonizing industry or long-haul trucking. Notably, one version of a hydrogen system involves blending hydrogen with natural gas, which could create lock-in to gas infrastructure. Gas with a bit of hydrogen can be treated as "lower carbon," though the gas can only be up to 15–20 percent hydrogen or so to work in existing pipes. Renewable natural gas—for example, waste methane from dairy farms—can also be blended in to make low-carbon gas. One key concern that community groups have with combusting hydrogen for power is emissions of nitrogen oxides, and long-term exposure to these can increase the risk of respiratory conditions. So even if greenhouse gas emissions are reduced by using hydrogen to provide reliable electricity, there still may be serious public health concerns.

However, if these approaches to decarbonizing fossil fuels work—and there's no technical reason why they won't—they totally change the politics around the aspiration of "ending fossil fuels." Many consumers don't care where their energy comes from. They just want it to be there. Fossil fuel companies can make a real case for being able to deliver reliable energy with a lower carbon footprint. The only reason they haven't decarbonized already is because no one made them.

The public spending in US legislation and elsewhere will bring down the cost of decarbonizing fossil fuels.

At this point, readers who are well-versed in energy systems and deep decarbonization may be thinking, given the alternatives of energy scarcity or climate change, what's so bad about having lower carbon fossil fuels as part of the future energy mix?

There are some critical reasons to transition away from fossil fuels even if they can be decarbonized. Climate change may not even be the most pressing reason to end fossil fuels. A narrow focus on climate change—seeing the world through carbon glasses—tends to parcel these reasons out from discussion. They range from air pollution and health impacts to the role of fossil fuels in supporting corrupt regimes, as discussed in chapter 5.

Making the Case for Phaseout in a World Captivated by Net Zero

When faced with the question of what's more realistic—a fossil fueled, high-emission version of net zero or a net zero integrated with a planned phaseout that takes residual emissions near to zero—the latter is the sensible policy objective. People know this, but in a fuzzy way. The details of how these worlds and policy pathways differ will become increasingly clear as we arouse ourselves from the net-zero hangover.

This book is for people who are not sure about net zero—people who are understandably confused about net zero, or people who know about net zero and think it would be a miracle if we got to it, so leave well enough alone. But it's also for people who are passionate about ending fossil fuels: it's filled with challenges that will help you sharpen your arguments about leaving oil, gas, and coal in the ground. Because when the promise of decarbonized or recyclable fossil fuels is extended as a way to keep enjoying things we are used to—and when such a promise is extended in an era of global anxiety and skepticism about the impacts of renewables—it will be a lot harder to argue for a complete end to fossil fuels. Anyone who wants to push for it will need a sophisticated and wide-ranging set of arguments to combat the reasoned, seductive logic of cleaner, recyclable fossil fuels.

That's why this book offers two primers in one: what you need to know about net zero, the basics on fossil fuel phaseout, and how they are linked. The first part offers a backgrounder on net zero: where it came from, what it implies, and why it's an inadequate framework for what we need to do. Net zero in a sense is boring (that's in fact part of why it has had such a robust career), but you'll know what you need to know about the science and history of it after reading the first chapter. In part 2, this book offers a framework for how we can think about the problem of phasing out fossil fuels. When phaseout is discussed in policy, it still ends up being a rather technocratic, vague, or nationally oriented discussion. This framework offers several different ways of thinking about phaseout: as a problem of culture, of international relations, of code, of infrastructure, and of building political power. In this section, we add some dimensions to our understanding of the challenge. Finally, the third part summarizes the policies that have been suggested to actually wind down fossil fuel production.

Even though this book engages with a lot of ideas around just transition, it's not strictly about energy transition—it's just about one part of this. "Energy transition" typically encompasses both the ramp-up of new energy sources and the exit from old ones. Obviously, ending fossil fuels without scaling up clean energy would be catastrophic. Moreover, the transition needed isn't just an energy transition, but a materials transition. Around 12 percent of global oil demand goes to petrochemical feedstocks. We wear clothes made from petrochemicals, package our goods in them, use them in medicine and cleaning, and drive on roads made from them.[30] Oil and gas-derived products are everywhere. A key strategy for the oil and gas industry's survival in a decarbonizing world will be the demand for petrochemicals; the IEA expects them to be the largest driver of demand for oil and gas in 2050. "We produce your athleisure wear and the chemicals that make your cold storage vaccines possible," will be the oil and gas industry's answer to ending oil and gas production. And this is one reason why oil and gas will not simply disappear. These transitions are inevitably entangled with the idea of a planned decline for fossil fuels, and avoiding a negotiation between fossil fuel companies and society where we simply accept more petrochemical products for less fossil fuels is key. This book, however, takes a narrow focus on the ending part—the deliberate decline of fossil fuels in particular, not oil and gas extraction broadly—because there is already a lot of good writing on how to ramp up renewables and restore ecosystems to create a cleaner and greener world. In contrast, writing on how we actually go about ending the old is comparatively scarce. Perhaps this is because the ending-of-things seems really hard, not to mention depressing.

Yet we can learn to think differently about the challenge. We need to, because the challenge isn't unique to fossil fuels or to this moment. Think about our tortuous relationship with exploitative information technology platforms, all the toxic practices in our food system, and unsustainable water extraction, to name just a few examples. There are all kinds of things to move away from: things we must stop doing because they are killing us; things we must retreat from because they are doomed, as a part of adaptation; and technologies that are foisted upon us which we can choose to reject. Policy-wise, these are all in completely different topical boxes, but culturally they are related to a basic capacity to end things. This is a capacity related to political power. But it's also a psychological capacity. And it's about planning. Ending things, in a planned fashion, is a capacity we need to develop in order to thrive. While this book is about fossil fuels and about urgent action items for this particular moment, it's also about this broader capacity of managed decline, managed retreat, managed rejection. These things aren't negative: they're fundamentally affirmative. They're about taking control of our own destiny.

Part I. The Cruel Optimism of "Net Zero"

1

How to Not Say the F Word

Net zero is both tedious and a hot trend. Net zero has the effect of being able to instantly transform the exciting, terrifying work of our times into something numb and boring. Net zero is the stuff of webinars, white papers, consultant reports. But it's also popular. At the time of this writing, at least six countries have a target of net-zero emissions by mid-century solidly in law, several others have them in proposed legislation, and another dozen have net-zero targets in policy documents. Scores of cities and subnational jurisdictions have also announced net-zero goals. Many of the world's largest companies, from tech giants like Apple to fossil fuel producers and utilities, have also jumped aboard the net-zero train. Microsoft aims to be not just net zero but carbon negative by 2030, and remove all of its historical emissions since its founding in 1975.

Even the Green New Deal, a conceptual innovation that links climate change with broader social goals, begins with the climate math. House Resolution 109—the resolution introduced in the House that serves as a sort of vision document —kicks off by reporting findings from the Intergovernmental Panel on Climate Change's Special Report on 1.5°C. Keeping temperatures below 1.5°C will require "(A) global reductions in greenhouse gas emissions from human sources of 40 to 60 percent from 2010 levels by 2030; and (B) net-zero global emissions by 2050," it states. The resolution talks about achieving "net-zero greenhouse gas emissions through a fair and just transition for all communities and workers."[1] It's a tremendous document. But it doesn't lead with ending fossil fuel production. In fact, it doesn't even mention the words "fossil" or "fuels."

Neither does the landmark Paris Agreement. Read the texts, and you'll be struck by this weird verbal jujitsu of documents aimed at ending this thing that they can't even name. The European Green Deal is not much better, either: it does address fossil fuels a few times and call for phasing out subsidies to fossil fuels, but it certainly does not center production.[2] Rather, these—and most climate policy—center emissions, the byproduct of combustion, thus skipping right past production.

This isn't a new observation. The Indigenous Environmental Network, for example, released a statement that strongly rejected the net-zero emissions language in the resolution. While they were grateful to see legislative support for climate action, they could not fully endorse the resolution until it made explicit the demand to keep fossil fuels in the ground. "Furthermore, as our communities who live on the frontline of the climate crisis have been saying for generations, the most impactful and direct way to address the problem is to keep fossil fuels in the ground," they wrote. "We can no longer leave any options for the fossil fuel industry to determine the economic and energy future of this country."[3]

While activists pointed out the omission of references to fossil fuels, the Green New Deal was still generally celebrated as an important step. The innovation of the Green New Deal is the way it broadens concern to include the "several related crises" the resolution goes on to list: declining life expectancy, income inequality, and racial and gender wealth and earnings gaps. But what starts out as a remedy for inequality may end up as a carbon accounting exercise.

Rather than jump into figuring out the nuances of a just supply-side climate policy—which is really hard!—well-meaning climate and sustainability professionals may spend several years wandering in the thickets of "net zero." Again, net zero, at its simplest, means balancing some amount of positive greenhouse gas emissions with negative emissions or removals. Well, what amount of positive greenhouse gas emissions are we talking about? That often remains conveniently unclear. Goldman Sachs, for example, estimates 25 percent of current emissions would remain. They state that these anthropocentric greenhouse gas emissions are not currently abatable using available large-scale commercial technologies, calling for both innovation and investment in sequestration technologies to achieve net zero.[4] A quarter of current emissions is still a lot of emissions. Decarbonization roadmaps set out by countries and cities tend to be vague on this point, when they even exist, but many of them include 10–20 percent of baseline emissions as "residual" emissions, or remaining positive emissions, which will need to be compensated for by removals. This is a looming political fight, though in some places, it has already been decided: New York State, for example, has set the maximum residual emissions at 15 percent of 1990 levels; Massachusetts law similarly requires an 85 percent reduction from 1990 levels.

Is net zero really more ambitious than what preceded it? Net-zero targets have replaced earlier targets, which often took nonintuitive goals, such as an 80 percent reduction of greenhouse gases from 1990 levels by 2050. Replacing these reduction goals with net-zero targets has generally been welcomed as a strengthening of ambition. But it may not be in all cases: in theory a country could have a net-zero target that is less than an 80 percent reduction of greenhouse gas emissions from some earlier point, as long as those positive emissions can be compensated for by negative ones.

Net zero by 2050 is an ambitious goal relative to the pace of climate action thus far. Achieving it would be incredible. But there are also ways in which it is the wrong goal. The concept of net zero offers balance and stability. It also offers ambiguity that can be exploited.

Cleaner Fossil World or Near-Zero World?

Imagine a world that has reached net zero near the end of this century. Let's call it "Cleaner Fossil World." Cleaner Fossil World still has fossil fuel companies, particularly oil and gas companies. They have shifted their portfolios and are now producing fossil fuels with lower carbon intensity. Oil is still produced for aviation, shipping, and industry. The companies have invested in systems to capture billions of tons of carbon dioxide from the atmosphere, injecting it underground and turning it into fuel, an energy intensive process. This allows a class of people in the developed world to keep internal combustion vehicles with drop-in fuels. Meanwhile, energy has become more expensive, as fossil fuel companies have passed the cost of all their new carbon-capturing equipment down to consumers, and many people live in energy poverty, carefully rationing the solar power they have access to.

But this Cleaner Fossil World has achieved net-zero emissions by sucking up tremendous amounts of carbon. Vast tree plantations that have decimated biodiversity provide a portion of those negative emissions. Land is routinely appropriated for carbon storage and renewable energy generation. The companies and platforms that finance, arrange, and perform the industrial services of removing all this carbon hold tremendous power, because people rely on them for climate stability. Cleaner Fossil World might align with the International Energy Agency's Sustainable Development Scenario or Shell's Sky scenario. In this world, there's a circular carbon economy and a continued role for oil and gas, but oil and gas production is decarbonized.

Now imagine a second world that has also reached net-zero emissions near the end of the century. In Near-Zero World, the remaining greenhouse gas emissions arise primarily from agriculture. The need for negative emissions is lower, and they can be generated through modest infrastructures. Near-Zero World might follow along with the International Energy Agency's Net Zero by 2050 scenario, which paints a world that quintuples investment in solar photovoltaic technology by 2030, shuts down most coal plants, electrifies half the vehicle fleet by 2030, retrofits buildings, and sees significant behavioral changes.

Cleaner Fossil World and Near-Zero World might sound pretty much the same. Anyone could be forgiven for thinking they are the same, if they heard about them casually. Both of these worlds have attained net zero. But which is more livable? Which is more plausible? What "net zero" does is allow companies and policymakers to conveniently ignore the choices between these worlds and countless others. Net zero may be a temporary state on the way toward a fossil-free future, or it may be a permanent condition where fossil fuels continue forever, re-interpreted as part of sustainable carbon management. Net zero does important work: it shifts attention entirely onto emissions, counting and balancing them. This draws attention away from the point of production, which is where we need to also be focusing.

The argument here is this: without deliberately phasing out fossil fuels, we're more likely to end up in Cleaner Fossil World. But that's not even all that likely. What's even more plausible is that the world never reaches net zero at all. The structures of power in place would only half-heartedly go about decarbonizing fossil fuels and call partway "good enough," leaving us with a devastating 3°C of warming. Conversely, the chances for a livable planet are much higher if we frame our goals around ending production, aiming for an actual fossil-free world this century. Arguments about ending fossil fuels tend to read as Puritan ones: no compromise, total abolition—and so they get laughed off by "serious folks" as ideological or unachievable. The set of arguments in this book come from a different rationale.

2

Inventing Net Zero

What is "net zero," and how did it get into policy ambitions? Whose ambitions are these, anyways? It seems like the "net" just kind of crept in, quietly, over the past few years and attached itself to zero, like a little invisible demon or imaginary friend. But like all concepts, it was designed by people in particular places with particular visions of the future.

Climate Home News editor Megan Darby traces the net zero concept to an informal network of women—climate activists, lawyers, and policy professionals—who met at a Scottish country estate in 2013 and discussed what would be a workable yet ambitious goal in the wake of the failures of the earlier Copenhagen climate summit.[1] Net-zero thinking made its most recent debut in the 2015 Paris Agreement, in which nations agreed on the goal of limiting warming to 2°C above preindustrial temperatures and striving to limit warming to 1.5°C. Key idea here: the Paris Agreement aims to do this by reducing emissions as soon as possible, "to achieve a balance between anthropogenic emissions by sources and removals by sinks of greenhouse gases in the second half of this century."[2]

The earth is a spinning ball of flows between atmosphere, soil, plants, oceans. Those flows have to be broken down into accountable numbers belonging to national territories in order to set up a regime of responsibility for managing them, as policy researchers Eva Lövbrand and Johannes Stripple chronicle. The United Nations Framework Convention on Climate Change negotiations in the 1990s began these accountings of emissions and removals.[3] Northern forest countries introduced net emissions logics into negotiations, which crystallized in the 1997 Kyoto Protocol agreement that opened up the possibility of using land-based carbon sinks to comply with their reduction targets.[4] The Kyoto Protocol embodied centralized, top-down, compliance-based logic.[5] The "national carbon sink" as a resource and idea was born. The net accounting system was part of the compromise of Kyoto, but it remained the subject of intense controversy. What counts as a forest? Can the amount of carbon in a forest accurately be measured? Does anyone actually want to pay to protect and grow forests?

Carbon cowboys, junk credits, land grabs: the chicanery, violence, and misery of the early carbon trading era has been well documented. There are two types of carbon markets: voluntary markets (supported by companies that decide, out of corporate sustainability or other motivations, to offset their emissions) and compliance markets, which are mandated by law under cap-and-trade schemes, such as in the European Union and California. The EU Emissions Trading Scheme experienced early significant failures, prompting questions about the success of markets. But the market rationale kept churning on in the private sector, and several voluntary carbon offset standards have evolved, some of which aim to treat earlier problems.

Today, both voluntary and compliance markets are rising. The EU's Emissions Trading System has seen carbon prices soar, hitting record highs in early 2021 of nearly 40 euros per ton of CO_2 equivalent (US$49). In 2020, carbon markets reached a value of 229 billion euros, which is a fivefold increase from 2017.[6] Is this an indicator of climate action going mainstream? Does it show that companies are speculating that emissions allowances will be constrained? It seems that is part of it. The belief in successful climate policy could even produce a rush for carbon futures, so that companies can lock in cheaper offsets or removals now, which, if it happened, could drive speculative land grabbing. The flip side of such a rush is that the money pouring into carbon futures could pay for the development of the technologies needed to remove carbon.

At the time of writing (April 2021), there's a renewed interest in both carbon offsets and carbon removals, as well as widespread confusion about the distinction. An offset involves credits from projects that *avoid* or *reduce* emissions. This involves proving that whatever action a project took avoided something that would have happened anyway—for example, if someone protected a forest, to get a carbon credit, they would have to prove that the forest was in danger of being cleared. For this reason, the offset market is not well set up to manage removals; too much of it focuses on calculating this counterfactual future. A removal, on the other hand, is capable of removing carbon already in the atmosphere. Some actors—including technology companies like Microsoft or Stripe—have recognized the problems with offsets and are moving away from them and trying to catalyze a market for carbon removals instead. All the companies that made net-zero commitments will be groping around for carbon removals. Right now, there is a reinvigorated enthusiasm for forest sinks, and virtually no supply for long-term carbon removals. But this is changing as big companies seek to partner with industrial projects that can permanently store carbon.

The Paris Agreement also brings the potential for a new international carbon market, where emissions reductions could be traded between countries. Article 6.2 in the Paris Agreement enables the transfer of "international mitigation outcomes," which offers the prospect of carbon unit trading, and Article 6.4 offers the ground for a Sustainable Development Mechanism. Article 6 has been called "something of a miracle" by Shell climate advisor David Hone, because there was no carbon pricing or carbon-market language in the draft text as late as the day before the final adoption of the agreement.[7] Article 6 is at least better than the previous situation under the Clean Development Mechanism, in that the Paris Agreement explicitly forbids double-counting. It also ushers in a new wave of quantification as nations define their Nationally Determined

Contributions (the pledges they make). So on a global governance scale, the international architecture has opened the way for a new system of trading carbon offsets as well as removals.

The vibe now in climate governance is described by words like "polycentric," "fragmented," "flexible," "networked," and "pluralist." But to understand why we seem to be back in the same place, with what sounds like the same type of approaches that haven't worked for the past three decades, we have to appreciate a basic feature of decarbonization: the capacity to decarbonize is unequal. For one, it's unequal geographically. If a country has forest resources that connote large removal capacity, and another country has none, should there be some way of trading?

Net zero, at its best, is the dream of a world in balance. Different nations have different capacities for how much they can decarbonize and how much carbon they will be able to remove. Not every nation can build an industrial carbon removal infrastructure, access hydropower, or plant carbon-sucking forests. Nations have different historical responsibilities and capabilities, too. A progressive version of net zero could be an instrument of climate justice. Imagine a better version of the United States, one that has built infrastructure for sucking carbon from the atmosphere and injecting it underground. This better version doesn't use that technology to compensate for the emissions of the highest payer, but rather uses it to compensate for nitrous oxide from fertilizer applications in sub-Saharan Africa, as one part of payback for climate damages. "The US would never do that." Probably not soon, but we should demand more. There's a version of net zero that could actually be an instrument of fairness and repair. The status quo version is that the US will develop giant carbon storage fields offshore and charge other countries for putting CO_2 in them, if it's even still capable of large infrastructure projects.

The capacity to decarbonize is also unequal when it comes to sectors. For years, analysts have been breaking down the decarbonization puzzle into pieces, looking at the capacity to decarbonize each sector. Most famously, Stephen Pacala and Robert Socolow at Princeton published a 2004 article in *Science*, as part of a project funded by Ford Motor Company and BP, which proposed fifteen "wedges" for decarbonization: each wedge was an action that could reduce emissions by 1 gigaton.[8] Wedges included things like increasing fuel economy for 2 billion cars from 30 to 60 miles per gallon, cutting carbon emissions by 25 percent in buildings and appliances, and so on: each sector could do its part, and together it would stabilize the climate. This became the dominant way of seeing the problem, and it is still dominant. Steven Davis and his colleagues write:

> The real and lasting potency of the wedge concept was in dividing the daunting problem of climate change into substantial but tractable portions of mitigation: Pacala and Socolow gave us a way to believe that the energy-carbon-climate problem was manageable.[9]

Not only is it manageable, as a collection of smaller problems, but perhaps these smaller problems are swapable, too —swap out one wedge of mitigation actions for a wedge of carbon removal. The logic fits seamlessly.

The wedge concept has been criticized for focusing on a series of technical fixes. Not only do these approaches not focus on social change or demand reductions, they generally fail to incorporate supply-side restrictions. Approaching the problem as a series of sectoral wedges also places key judgments firmly in the hands of experts, who understand the technical aspects of each sector, build the models, and thus define feasibility.

The allure of net zero remains: it offers flexibility to decarbonize what is easiest and to compensate for what is impossible or too challenging to decarbonize. But this brings up the question: Who has the power to define what is difficult to decarbonize?

3

What's Truly "Hard to Decarbonize"?

Net zero is supposed to allow for "hard to decarbonize" sectors to keep emitting for a while. But what's actually hard to decarbonize, and who gets to decide?

"I think the challenge is that this isn't well-defined. And so its fuzziness is used politically, or I don't know, quasi-politically in scientific spaces ... And that's when it gets a little bit worrying," Andrew Bergman told me. Bergman is an engineer doing a PhD at Harvard, and I talked with him to understand what the science actually says about what's hard to decarbonize. Bergman recently worked with a team of forty other scientists on an open-source carbon dioxide removal primer. One of the sections he worked on analyzes the scale of hard-to-avoid emissions and the carbon dioxide removal needed to balance them out. His team's number is a much smaller number than in other analyses, and I wanted to ask him why.

Estimates of carbon dioxide removal typically rely on integrated assessment models, which use economic, ecological, and technological assumptions to model mitigation trajectories. Most of these models calculate that to keep warming below 1.5°C, between 5 and 15 gigatons of carbon dioxide removal would be needed by 2050. Some of this is to compensate for ongoing emissions, but some scenarios also overshoot their temperature or emissions targets and then use carbon dioxide removal later in the century to make up for it. The problem with this whole thing is that the models aren't looking at what's actually hard to decarbonize. They're looking at what's cheap to decarbonize based on a bunch of economic assumptions. As

Bergman told me in March 2021: "It's not just engineering, it's often economic and it's based on our capitalist mode of transacting. Which is to say certain things are expensive right now, but that's because we have a sort of apparatus around them that makes them only work if it'll be profitable, obviously with a crap load of caveats there. But it doesn't have to be that way." In other words, assuming we could make social changes to what's expensive through policy—what kind of infrastructure would we need to compensate for just those emissions that are truly technically difficult or impossible to eliminate?

Technically Difficult to Eliminate

Bergman and team came up with a figure of 1.5–3.1 gigatons of residual greenhouse gas emissions, defining hard-to-avoid emissions "narrowly as emissions that will be either unacceptable to avoid from a social justice perspective or extremely physically difficult to eliminate within the given timeframe." That figure may strike you as very low, or disappointingly high, if you were hoping that this version of Near-Zero World was even nearer to zero. But it's a very big deal that a scientific team can come up with a far lower number of leftover emissions if they imagine a society willing to use policy to change the economics of these decarbonization technologies, and make that social imagination the starting assumption for the questions they are asking.

Let's take a look at where these last 1.5–3.1 gigatons of emissions come from.

Other Greenhouse Gases Related to Agriculture

Most of the leftover emissions in their analysis weren't even from carbon dioxide. Bergman explains: "We said, 0.8 to 1.9 gigaton CO_2 equivalent nitrous oxide will continue to be emitted, at least for the foreseeable future. Some of it from waste, but the vast majority from fertilizer evaporation, for basically feeding the world. And so that is the biggest source." Nitrous oxide comes largely from agriculture, mainly from overusing fertilizers and manure. These emissions can be abated by more targeted fertilizer applications, but it will be a challenge to eliminate them completely.

More crucial background: carbon dioxide from fossil fuels and industrial processes only makes up about 65 percent of global greenhouse gas emissions. Another 11 percent of greenhouse gas emissions comes from forestry and other land uses. Halting these emissions, from ending deforestation and instead planting new forests, is implied in net-zero discourse. The remaining greenhouse gas emissions are from methane (16 percent), nitrous oxide (6 percent), and F-gases (2 percent). Methane is a greenhouse gas that is more potent than CO_2, but it breaks down far more quickly—carbon dioxide lasts for several hundreds of years, nitrous oxide for 110 years, and methane for just around 12 years. Methane emissions come from natural resources (like wildfires and peatlands), agriculture, and industrial emissions, including leaks from pipelines. Methane emissions from agriculture, such as rice production and particularly from livestock, are tougher to eliminate. "Enteric fermentation" is a natural digestive process of ruminants (cattle, sheep, and goats), and comprises 30–40 percent of agriculture-related methane emissions. Methane, however, wasn't included in Bergman and colleagues' "hard to decarbonize" bucket, because of this shorter life span. It doesn't accumulate in the atmosphere in the same way that carbon dioxide does, and so constant methane emissions don't need to be continually offset using CDR. "While halting methane emissions would have a one-time cooling effect, continued, constant year-after-year emissions are balanced by methane degradation and don't lead to additional warming," Bergman explained.

Transportation

The number two source of leftover emissions in Bergman's analysis? Transportation, specifically long-haul aviation at 0.7 gigatons, and shipping, at between zero and 0.3 gigatons. "It really doesn't seem like anyone believes that aviation will be able to be fully decarbonized because of battery weight," Bergman explained. Liquid hydrocarbon fuels have high levels of energy per volume and per weight, which makes them good for transporting large volumes of goods or people. Even if low-carbon or zero-carbon fuels displace the fuels we use now, their combustion still generates hard-to-avoid emissions.

Let's talk first about aviation: short flights on small planes could be electrified, and perhaps a network of short flights could be created. A crop of startups aspire to do this. But long-haul flights will require zero-carbon fuels (such as hydrogen and ammonia fuels, biofuels, synthetic hydrocarbons, and solar fuels). When are those actually carbon neutral? Synthetic hydrocarbons require removing carbon from the atmosphere (for example, using direct air capture machines powered by solar or wind) and then reacting that carbon with dihydrogen from electrolysis of water. Creating these synthetic fuels in carbon-neutral ways requires an infrastructure of its own. As Shell advisor David Hone writes:

> our (non-biofuels) solar world might need to see the construction of at least 100 large-scale hydrocarbon synthesis plants, together with air extraction facilities, energy storage for night time operation of the reactors and huge solar arrays. This could meet all future aviation needs and be capable of producing lighter and heavier hydrocarbons for various other applications where electricity is not an easy option (e.g. chemical feedstock and heavy marine fuels). The investment would certainly run into trillions of dollars and take decades to implement.[1]

The challenge of zero-carbon fuels is why the aviation industry has been developing an offsetting scheme, the Carbon Offsetting and Reduction Scheme for International Aviation. It's also why United Airlines has invested in a direct air capture plant in Texas: United's CEO has noted that the math doesn't work out without direct air capture.

Shipping comprises close to 3 percent of global greenhouse gas emissions, and this industry too imagines a future with zero-emissions fuels. Governments at the UN's International Maritime Organization agreed in 2018 to halve their emissions by 2050 and to try to phase them out entirely, though it's not clear how that will be accomplished. Aside from the zero-emissions fuels cataloged above, ships can be designed to be more hydrodynamic and efficient.

There are one billion cars on the roads today. Bergman and colleagues envisioned full electrification, but this would be a major feat to do quickly. The average age of a car on the road is 7–8 years; the lifetime is 15 years.[2] Those existing cars equal another 10 billion tons of CO_2 emissions. Sales need to be electric vehicle–only by the 2030s to turn over the fleet to electric and meet midcentury decarbonization targets. Many jurisdictions are following the science on announcing phaseouts, with a wave of phaseout targets identified in 2016–17: Norway, Germany, India, Ireland, France, the United Kingdom, Scotland, China, California, Netherlands, Sweden. All that is great. But the old cars linger—cars are getting safer and lasting longer. So it's not just a matter of stopping production but also dealing with what's already here. Retiring the fleet could be hastened by buying back cars with internal combustion engines, but it would cost half a trillion to deal with the 110–125 million internal combustion engines on the road in the United States.[3] Looking at the options for decarbonizing transportation, you can see why some in the climate policy community would throw up their hands and say, full decarbonization isn't feasible in this timeframe, so we need to compensate these continued emissions with negative emissions until the internal combustion engines have naturally faded out.

Technically Possible But Still Really Challenging

In Bergman and colleagues' analysis, then, agriculture and transportation are the technically difficult emissions to eliminate. Two other sectors are typically classified as hard to abate: firm electricity and industry. Bergman assumes that we could fully decarbonize these, if we decided to spend our resources and effort on it. But we should briefly review why other analysts throw them in the hard-to-decarbonize box.

Electricity

Solar energy has become astoundingly cheap. It's now cheaper than new coal or gas-fired power plants. Solar energy has grown fortyfold globally from 2008 to 2018.[4] And it's been a sudden revolution: more than 99.9 percent of all photovoltaic modules and concentrated solar plants ever built were installed after 2008.[5] Renewables are on track to meet 80 percent of the growth in global electricity demand to 2030, according to the International Energy Agency, meaning that even with the fairly mediocre climate policies we have now, there isn't that much justification for building new fossil fuel infrastructure. The costs of solar photovoltaic modules have fallen by more than a factor of 10,000 since they were first sold on the market.[6] Experts and models originally underpredicted the success of solar; according to current expert forecasts, it's supposed to set new records for deployment each year.[7] It's easy to be impressed with solar's performance—but it still supplies just 1–2 percent of global energy.

Still, when you see growth numbers like this, you might think, at least we've got this part of it figured out. Decarbonizing electricity is the part we know how to do—to a point. US emissions, for example, peaked in 2005, largely because of replacing older coal plants with cheap gas and renewables. However, the power sector emissions reductions we've already seen are regarded as the cheap, easy, low-hanging fruit. There are still plenty of gains to be made: it's generally estimated that a 50–70 percent reduction in carbon emissions from electric power can be achieved by technology that's commercially available today.[8] It's that last 30 percent that's the problem.

There are a few challenges to decarbonizing the power sector all the way in the timeframes we need to. You've probably met That Guy Who Always Asks, "What happens when the sun doesn't shine, or wind doesn't blow?" Enter the debates about "highly reliable" or "firm" electricity, electricity that can deal with variability in both demand and output. Is "unfirm" or "unreliable" just a slur that anti-renewables people like to hurl? Not really: having an unreliable grid can be deadly if there's a heat wave. Fluctuations with wind and solar are day-to-day, minute-to-minute (passing clouds), and seasonal. The "capacity factor" of wind or solar refers to the percentage of average working hours of a power plant in a year—for example, a wind farm may work 1,800–2,300 hours in a year, while a solar power generator may work 800–1,200 hours. So wind would have a capacity factor of 20–50 percent, while a solar power plant may have one between 10 and 30 percent. Intermittent sources need backup capacity to guarantee availability.[9]

There are certainly ways to deal with these periods of unavailability. Building more wind and solar than is actually needed ("overbuilding") is one. Transmitting power from other locations is another. Countries can build a larger grid with more transmission lines that connect spots where wind and solar are stronger and more constant, in order to send power from sunny places to less sunny ones; perhaps a "supergrid" that connects countries or even the whole globe. There's also a number of forms of storing energy for later. Water can be pumped into reservoirs for later release and generation of hydroelectric energy ("pumped storage"). Electricity can also be stored by compressing air in geologic formations and then recovering it with turbines when it is released. Electric vehicles could be used to support the electrical grid with vehicle-to-grid technologies, which store spare power in everyone's car batteries in a kind of distributed storage network. Thermal storage systems can store heat in water tanks, buildings, or solid materials. There are also demand-side management techniques, like interruptible supply contracts, or demand response relating to smart thermostats. But electricity storage is one area where there are actual technical barriers.

What about other options for clean, firm energy? Firm zero-carbon energy can also come from flexibly operated nuclear power plants, hydropower, fossil fuel plants with carbon capture and storage, geothermal, and biomass. None is perfect: hydropower and geothermal are limited by geography. On-demand gas plants with carbon capture and sequestration (CCS) still involve production of fossil fuels; biomass can demand land as well as compete with the need to use biomass for liquid fuels. It is going to be hard, though, to argue for eliminating the option of gas with CCS: it will be portrayed as playing with the security of people's power supplies. In the gas battles of the 2020s, the net-zero grid will be described with words like "flexible" and "inclusive" compared to the carbon-free grid.

The retirement of nuclear plants is making the situation trickier. Many nations, like Germany and Japan, have been decommissioning nuclear power plants; other places are letting them retire at the end of their operating life. Nuclear used to provide 17.5 percent of the world's total electricity generation, at its peak in 1996, but now it provides just 10 percent.[10] Readers may be thinking—good. But nuclear has a high capacity factor—a plant would be online more than 90 percent of the time, compared with a coal or gas plant that might be online half the time, because of refueling and maintenance.

Moreover, people often don't appreciate how much power nuclear power plants provide. A typical nuclear plant would provide one gigawatt of power. What's one gigawatt, anyway? Over 3 million solar panels, for example, or 400 wind turbines, both of which are significant in terms of the land they occupy.

The danger of phasing out nuclear without having something else online means that fossil fuels may be called on to fill the gap: this is why Germany's CO_2 emissions went up again in 2017 after years of hard-won declines. The nuclear plants were substituted with brown coal. What about building new nuclear plants? Right now, there are about 413 nuclear reactors in thirty-one countries, dominated by the United States, France, China, Russia, and South Korea, countries that generate 70 percent of the global total of nuclear power.[11] There are some new nuclear reactors being built—particularly in China, and with new countries entering, such as Bangladesh, Belarus, Turkey, and the United Arab Emirates—but many of these are behind schedule. Because these projects have been so slow and costly, scientists are considering whether small modular reactors would be more cost-competitive. These are reactors designed with modular technology, generally 300 megawatts electric or less. But they face a paradox—they would need a lot of orders to be placed to get to the point where creating an assembly line delivers the cost benefits from the modular design; custom building each one doesn't deliver those benefits. But who's going to want to order hundreds of these when they haven't been proven yet?

Back to the challenge of decarbonizing electricity: alongside the issues of firm power and nuclear phaseouts, there's a question of the raw materials and land needs of scaling up renewables. This is a materials challenge. A World Bank analysis of mineral demand for renewable power and energy storage indicated that 200 million tons of iron, 80 million tons of aluminum, and 30 million tons of copper could be required for wind, solar, and battery storage through 2050 in a 2°C scenario.[12] A review of the literature on minerals for renewables found that this unprecedented extractive industry could in some cases exceed current reserves by 2050, and it's also hard to know what the emissions would be, because greenhouse gas emissions from all that mining are not at all transparent.[13] While this isn't the main reason firm electricity is "hard to decarbonize," it is worth keeping in mind given that the mandate to "electrify everything" is expected to grow demand for electricity significantly, and material use will grow with it.

Industry

Industry is hard to decarbonize for two reasons. First, "process" emissions arise directly from industrial processes. Second, some industrial processes need high-temperature heat. Roughly 10 percent of global greenhouse gas emissions come from burning fossil fuels to generate heat for industrial processes. These processes include chemical conversions, glassmaking, refining ores. Calcination of limestone for cement requires temperatures of ~2500°F, melting iron ore to produce steel requires 2200°F, and steam cracking to produce ethylene, a feedstock for plastics and petrochemical, requires 1500°F.[14]

What are the options for decarbonizing industrial heat? One is simply to continue to burn fossil fuels but add postcombustion CCS technology to scrub out the CO_2 emissions. Sometimes carbon capture and storage gets called a "false solution," but putting it onto cement plants or other industrial facilities is a really good use of the technology. Cleaner heat sources could also be biofuels, hydrogen, electrified heating (with a clean grid), or nuclear reactors. Most of these approaches are far from commercial. Options beyond blue hydrogen or fossil fuels with CCS can double the cost.[15] Here "hard-to-decarbonize" seems to mean expensive. But the technical capacities exist. Some firms are looking to enter this space, such as the Gupta Family Group Alliance, a British-based international business group that aims to be the world's first carbon neutral steel company by 2030. It aims to use electric arc furnaces powered by renewables to recycle scrap steel, rather than make primary steel using a blast furnace; it also holds investments in renewables and is building green hydrogen steel plants. ArcelorMittal is also developing green hydrogen for steelmaking and offered the first verified green steel for customers in 2020. These may be boutique investments for companies that have a lot of capital to invest, and they are doing discursive work to reassure investors and maintain social legitimacy. But they are not doing this alone—the European Union and other governments are funding innovation in industrial decarbonization.

Decarbonizing something like steel or petrochemicals is especially hard because these are traded on global markets, and so a country attempting to decarbonize production could quickly become uncompetitive. Border tariff adjustments could help. But these are different industries than others: they're not consumer facing (when was the last time you went shopping for steel?), they're politically protected, and they're often treated as national security concerns. The vast difference in cost for decarbonized industrial products highlights the need for coordinated action. If one determined country undertook this transformation, some level of production would just shift elsewhere.

What about Degrowth?

Looking at various activities on a sectoral basis, as well as through the lens of what is technically "hard" to decarbonize, depoliticizes the choices around the way forward. While it is true that there are biophysical aspects that make decarbonization of various activities more difficult than others, at present there is no standard definition of "difficulty." In the real world, these distinctions of "difficulty" will be political and economic, not just technical.

Consider the hypothetical plight of Modest Co-Op, which for the sake of argument is a worker-owned factory in a rural area where there aren't many jobs. The sectoral logic puts Modest Co-Op in the same situation as Big Corp, which may be a conglomerate that is also polluting the local waterways and is abusive to underpaid migrant workers. Both of these enterprises might have high emissions from running their machinery. Now imagine a net-zero policy that says there can still be high levels of emissions—you just have to pay for all the corresponding removals. Big Corp will be able to afford this because it is screwing over workers and the environment, and it's also more likely to be able to invest in fancy new equipment. Modest Co-Op will go out of business, giving Big Corp even more power. What's an alternative? The government could help every factory get its emissions as close to zero as possible by offering subsidies for cleaner energy and new equipment. This isn't super radical—but the point is that there are big differences here for what "net zero" might mean for a community. We have to keep engaging with these details.

Looking at this matter of "hard-to-decarbonize" also naturally brings us toward conversations of degrowth. What about #flyingless, #shippingless, and #drivingless? Eating less meat? Using less steel? Clearly, some degree of this is needed—and this is where a sectoral approach outlives its utility, because driving less isn't purely a consumer choice. It requires changes to housing patterns, working patterns, parking policy, support for cycling and transit, and more, so that it becomes a viable option for people. Pretty soon the neat sectoral demarcations bring us to an everything-project.

There are certainly specific high-income populations who need to x, y, and z less, but then, there are many others who need energy access. A contraction and convergence approach to degrowth and growth is one equitable way forward. Shell's approach in their report *A Better Life with a Healthy Planet: Pathways to Net-Zero Emissions* actually implies such an approach, noting that people in the United States consume 300 gigajoules of energy per person per year, Europe 150, and

China 100. One hundred gigajoules per person per year "is approximately what is required to fuel the energy-based services that support the decent quality of life to which people naturally aspire."[16] (For reference, a gigajoule is a billion joules; a Paris to Singapore return flight is about 100 gigajoules per passenger.) They note that if everyone used 100 gigajoules per year, the current energy system would need to be doubled (from the current 500 exajoules to 1,000 exajoules by 2050). I am not suggesting that the company culpable for massive devastation in the Niger Delta is necessarily the best model for equity or determining how much energy a person needs for a healthy life. But I am suggesting that their numbers and logic be scrutinized and engaged with, because unless we calculate different numbers, those will be the ones on the table for discussion. And frankly, reducing US energy use to 100 gigajoules per person while doubling the world's energy production seems like a reasonable starting point—Shell's put forth a far more radical proposition here than I've heard from most quarters, and ironically, it could be read as far to the left of most of the Democratic party. But it's clear that a collapsed notion of "degrowth" doesn't quite describe the work here.

This book is focused primarily on supply-side reductions of fossil fuels. Readers interested in demand-side reductions and consumption, or seeking a more robust and nuanced discussion of degrowth, may want to check out the dialogic *In Defense of Degrowth: Opinions and Minifestos* by Giorgos Kallis and others; *The Case for Degrowth*, by Giorgos Kallis, Susan Paulson, Giacomo D'Alisa and Federico Demaria; Leigh Phillips's *Austerity Ecology and the Collapse Porn Addicts: In Defense of Growth, Progress, Industry and Stuff;* Jason Hickel's *Less is More: How Degrowth will Save the World*; and Kate Soper's *Post-Growth Living: For an Alternative Hedonism.*[17]

There are certain activities where reducing consumption or demand makes a lot of sense, and others where it doesn't. I would not want to paint eliminating junk mail or spam email with the same conceptual brush as eliminating flying, because aviation does a lot of good—but nor would I want to put the millionaire with a private jet in the same category as a climate migrant who wants to fly to visit family in another country. Discernment is important here. If we develop the social, cultural, and political capacities to phase things down, we can apply that capacity to degrow certain things that we decide are not socially valuable. Embracing planned phasedown will help us to degrow certain things, starting with things that are obviously wasteful, harmful, or annoying, like single-use plastics or gas-powered leaf blowers. Successful phaseouts of these can act as models to encourage a more positive view of phaseout and help create a positive feedback loop. Planning for phaseout is a way of getting at the tricky question of *how* to begin actually doing degrowth in a targeted way. Planned phaseout is essentially a technique of degrowth.

4

Creating Negative Emissions

For every ton of emissions deemed too hard to eliminate, there will be a theoretical ton of negative emissions to reach for. But do they really exist? Many technologies and practices are available for removing carbon from the atmosphere, though none are easy to scale up to levels that will make a difference in the climate.

Carbon can be removed through planting forests or farming in ways that store more carbon in soils. "Natural climate solutions" have been championed not just by conservation NGOs like The Nature Conservancy or the International Union for Conservation of Nature, but by companies and countries, including industries like fossil fuels and aviation that know they will need them. The fossil fuel industry's attempt to get together and figure this out, the Oil and Gas Climate Initiative (OGCI), is a group with members like BP, Chevron, ExxonMobil, Royal Dutch Shell, Aramco, and China National Petroleum Corporation. OGCI has several objectives: reducing methane emissions from oil and gas; lowering the carbon intensity of their fuels; scaling up carbon capture, utilization and storage; exploring whether they can play a role in the potential of emerging carbon removal technologies; and "actively exploring measures to catalyze natural climate solutions at scale."[1] Natural climate solutions offer something for everyone. And they are relatively affordable: Natural climate solutions could provide 37 percent of the cost-effective climate mitigation needed through 2030 for a 66 percent chance of curbing warming to 2°C, and a third of this can be delivered at less than $10 per ton, according to a scientific analysis that examined reforestation, biochar, conservation agriculture, improved grazing, coastal restoration, and more.[2]

Scaling up afforestation could be beautiful. The same goes for soil carbon sequestration—using methods like no-till, crop rotation, and/or regenerative grazing to store more carbon in soils. Agroforestry, or growing crops along with trees, can be a win-win for farmers seeking to diversify their crops. Blue carbon, or sequestering carbon in coastal ecosystems like seagrasses and mangroves, could double as coastal protection for communities and support adaptation.

These strategies to remove carbon could also be dystopian, depending on who goes about them and how. Imagine a network of satellites and sensors feeding data to platforms 24/7, optimizing each square meter of land to sequester carbon. That could be great—but now imagine it's run by a mega platform whose main aim is to allocate lowest cost carbon removals to algorithmic bidders. Companies procure carbon removals as needed based upon their changing forecasts, and speculators also exchange carbon removal futures, reserving vast areas of land from food production. The land is optimized for carbon, not grasshoppers, birds, or salamanders. Formerly forest-dwelling communities have long since left or been forced away, migrating to megacities, their cultural practices threatened with erasure. Despite early gestures toward vague "co-benefits," the discourse of climate emergency has led the lands of the world to be optimized for a sole purpose. This is one extreme end-of-the-century scenario, but it could be the direction that carbon logics + artificial intelligence + a state of climate emergency + a lack of socioecological systems thinking points us toward.

Why Industrial Carbon Removal Is Required

Whatever the implementation, there are three main problems with relying on ecosystems to deliver negative emissions. First, there are limitations on land. The most ambitious calculations of afforestation capacity rely on switching to plant-based diets and afforesting land currently grazed by cows—which may not be feasible or just around the world. Second, ecosystem-based carbon removals are finite in terms of how much carbon they can store over time. These carbon sinks, once filled, plateau in terms of what they can continue to remove. Third, ecosystem-based methods are vulnerable to climate change itself. Unchecked climate change could flip forest carbon sinks into carbon emitters by midcentury.[3] Given that nature has a limited capacity to absorb our carbon waste, other players—policymakers, scientists, and companies—are eyeing ways to store carbon in long-lived products or underground. This is expected to be especially important later in the century. Two main approaches are biomass carbon removal and storage and direct air capture with storage. These approaches already exist. They build on carbon capture and storage, which has been in use for decades; there are about twenty carbon capture and storage facilities worldwide, despite the low investment in the pollution control technology.

Biomass carbon removal and storage involves using biomass to store CO_2 underground or in long-lived products. The biomass might come from crop wastes and residues, managed forests, algae, or seaweed. After harvest, it is converted to products or fuels, in which case the CO_2 is captured and stored underground. Direct air capture involves building industrial systems that suck carbon from the ambient air. One design uses water solutions with hydroxide sorbents, which requires high-temperature heat to regenerate the solutions; the other approach uses amine materials bonded to a solid support and can use waste heat. Again, the carbon then has to be transported and injected underground for this to permanently remove carbon. There are ways of doing this, but they are relatively expensive—one direct air capture demonstration plant does the process at around $600 per ton of CO_2, but companies and analysts think that it can get in the $100–$200 per ton range. If successfully scaled up, direct air capture would pretty much set the price for carbon. Direct air capture requires copious amounts of renewable energy if it is going to be carbon negative. For example, one analysis found that removing the majority of current emissions, 30 gigaton (Gt)/year, would require around 50 exajoules (EJ)/year of electricity by 2100, and 250 EJ/year of heat, representing over half of today's final energy consumption globally.[4] Now, few people think that anyone would try to build out that much direct air capture—even removing 1 Gt/year would be quite a feat; 30 would be nearly unthinkable—but the point is worth noting. Climate-significant use of direct air capture assumes cheap and abundant renewable energy.

Every single one of these carbon removal technologies and practices has a spatiality that constrains its potential once it moves from models to the real world. Direct air capture in theory can be placed anywhere, but it is best done where there is ample renewable energy overlapping with favorable geology for sequestration. Biomass has a particular area where it can be sourced from and still be called carbon neutral. Oceans are not empty space for cultivating biomass but are widely used by coastal communities for fishing as well as farther offshore for shipping, as well as by other species, as efforts to set up marine protected areas have illustrated. The next few years will produce a bounty of spatially explicit analysis of where carbon removal can be done, but this may provoke a corresponding wave of despair about the prospects of reaching the scales that are projected to be needed to compensate for all those difficult-to-eliminate emissions.

The importance of these technologies is not just that they

could compensate for residual emissions. Industrial carbon removal technologies are important because they can remove legacy carbon from the atmosphere. Imagine every polluting country removing its emissions down to when the industrial era began, and thus lowering greenhouse concentrations back to 400 or 350 or 280 parts per million, removing some of the risks from climate change. There is a powerful climate justice case for doing this.

Obviously, that progressive version of carbon removal for climate repair is a very different thing than sucking carbon out of the atmosphere so that fossil fuel companies can continue producing fossil fuels. And at present, the expertise and technology for industrial carbon removal and geologic sequestration lies with fossil fuel companies. They have technical expertise in drilling wells, injecting fluids, laying pipelines, monitoring pressure, and all the rest. They also have institutional expertise in supply chains and financing large projects, as well as capital.

Looming behind the imaginary of carbon removal is the question: Can it actually be done? Is the world going to build out a whole second industry for putting carbon back underground and move a trillion tons of carbon from the atmosphere into the geosphere? Or is this purely a fantasy to make people feel better about continued extraction? What is really more realistic?

Net-zero as Collective Delusion

Think about how hard it will be to scale negative emissions: the financing, the infrastructure, the land, the social acceptance. Then think about the history of what's happened so far —the biofuel boom and bust, contributing to a global rush for land and food price spikes. The carbon markets that came and failed. The continued efforts at misinformation by the fossil fuel industry. The claims of making carbon recyclable atop the failures to actually recycle other forms of waste. How on earth could anyone think betting on our capacity to deliver gigaton-scale removals in a few decades would be reasonable? Does putting it on a graph automatically make it seem reasonable? Are all the climate professionals radical optimists? Is Twitter a technological enabler of collective delusion?

Climate professionals may indeed be optimists. Even if the vast majority of people involved in a collective delusion are kind, well-meaning people, the effect of that collective delusion can be completely sinister. Climate professionals are optimists raised with a sense of feasibility that is socialized to be expansive in some regards and constrained in others. They have *learned* that a sectoral approach that focuses on emissions reductions and is aimed at net zero is the way to think about this problem. Well, who taught them that? In the United States, many climate professionals were trained at Ivy League schools. There is a certain culture and set of social norms, and for the ones who got the coveted jobs in this sector, the system has largely been working for them. Just as salient may be who comprises the philanthropies and other institutions that fund these organizations and what their theory of change is. Both life experience and training can bear on a person's assumptions about how the world works and their implicit theory of change, which shapes how one bounds what's feasible or unfeasible. When it comes to this specific problem of decarbonization, thinking has been shaped by research at high-profile institutions which the fossil fuel industry has had a part in funding, as well as by research at institutions like the International Energy Agency, and by the scenarios work done by companies like Shell. Basically: what's feasible is conditioned by assumptions about several things, including (a) markets, (b) the role of the public and public power, (c) technical change, and (d) what governments can to do support change. All these assumptions are broader ones beyond the framing of this particular challenge—they are shaped by education, class, social networks, and experience.

The contemporary obsession with metrification, accounting, and modeling is part of what gave rise to the net-zero framing. (It's also part of why climate professionals hold power—the accounting is so complex it requires a class of professionals to deal with it.) Modeling is a legitimate form of storytelling about a collective delusion that masks its weirdness. There are two simultaneous things we need to do to address this. One is to deprofessionalize models and make them tools that everyone can use. The other is to argue for ways of seeing the problem that don't have to do with models and metrics.

Professionalization more generally is also part of the problem. By "climate professionals," I mean a class of workers that is employed in things like greenhouse gas accounting, branding, and campaigns, in both the NGO and private sectors. Environmental NGOs grew up in the 1970s and 1980s to become large-budget organizations in the 1990s; conservation NGOs in particular had not just employees but property.[5] With professional employees and assets came responsibilities to account for spending and take care of employees, and a level of bureaucracy ensued. There's been talk in many sectors about "NGOization"; this is true in the climate space, too. NGOs are defining policy and key terms—playing a leading role, especially given the decline of the environmental state. "Nature-based solutions" and "natural climate solutions" were basically invented by NGOs.

NGOs are restrained in some ways by their structures. Employees' livelihoods and continued funding depends on talking about the problem in certain ways, providing "deliverables" with data, making themselves indispensable to the solutions. But climate movements have helped shift the Overton window on what's possible with regards to climate action. I use this concept with some reservations—Joseph P. Overton was an executive at a free-market think tank. He named the idea that it's up to organizations like think tanks to propose ideas outside the window of what's considered possible, and to stretch this window. This idea reached a wider audience in Fox News personality Glenn Beck's political thriller *The Overton Window*, in which the protagonists discover a PowerPoint presentation on an elite scheme to reform education, energy, infrastructure, and even cap-and-trade. Laura Marsh, writing in *The New Republic*, chronicles the Overton Window's drawbacks as a theory of change, pointing out that "viewing politics through the Overton Window reinforces liberal notions about the moderate center, even as that center ground erodes."[6] Ben Burgis has a great critique in *Jacobin*, pointing out that demanding radical things can have a backlash effect and not even get you to some half-way solution—and shifting public opinion is not a substitute for building political power.[7]

When it comes to managed decline, though, stretching the bounds of the acceptable and building political power are part of the same work, and many climate professionals are just waiting for license to be more radical. The window of the possible *is* shifting. In a campaign debate between 2020 presidential candidates Donald Trump and Joe Biden, Biden stated he would "transition from the oil industry." Trump tweeted:

> Joe Biden made perhaps the most shocking admission ever uttered in the history of presidential debates. On live television, Joe Biden confirmed his plan to ABOLISH the entire U.S. Oil Industry—that means NO fracking, NO jobs, and NO energy for Pennsylvania Families!

But even having this discussion showed that it is a discussion that can be had. A planned ending to fossil fuels is becoming more normal every day; it just needs a bit more encouragement. This means repeating key policy ideas as if they are obvious and doable, and engaging in discussion about their drawbacks, all the while creating new social norms about winding down fossil fuel production. Let's look at a variety of different arguments that could help make the case.

5

Why We Need a Planned Ending for Fossil Fuels

Which sounds better to you?

a. Managed decline of fossil fuels
b. Energy transition
c. Phaseout of fossil fuels

"Managed decline" has the double-whammy of "management" (boring) and "decline" (grim).

Energy transition sounds less depressing, doesn't it? "Energy transition," unfortunately, is language that swaddles us in ambiguity (much like net zero). It doesn't explicitly state an end to fossil fuels. Rather, "energy transition" leaves open the possibility that renewables will come down in price, and this will naturally force a shift to cheaper energy sources: the old will be outcompeted and fade away.

I use the framework of phaseout in this book, because we urgently need to deliberately phase down fossil fuels as well as ramp up alternatives to fossil fuels—*and* we need to do both of these in a synchronized kind of dance, making sure that the clean energy is coming online fast enough so that no one is left without energy. This complex set of maneuvers is alluded to in the just transition discourse. In some instances, such as the Climate Justice Alliance's conceptual framework, just transition encompasses both growing the new and "starving" the old. But in other instances, "just transition," like "energy transition" more broadly, is an elastic concept that tends to focus more on scaling up the new.

The scientific community agrees on the climate math of ramping down fossil fuels, which offers the basic contours of the problem. To keep warming below 2°C, there is a carbon budget of around 1,100 gigatons (Gt; a gigaton is a billion tons). Burning this 1,100 Gt would get us 2°C of warming. But the greenhouse gas emissions contained in global fossil fuel reserves are around three times more than this amount. Thus, a third of oil reserves, half of gas reserves, and more than 80 percent of coal reserves are "unburnable" if we want to limit warming to 2°C.[1]

With the dominant way of thinking about climate math, the solution is expressed in terms of the right price—the carbon price that is high enough to deliver the replacements quickly and price out fossil fuels. When the math problem becomes one of finding the right price, this circumvents discussions of managed decline. One obvious problem with this is that jurisdictions may go out of their way to keep fossil fuel powered plants going even if they are unprofitable because of other reasons, such as jobs, pressure from unions, or pressure from corporate donors. Actors that pretend to be all about math, like big companies and rational policymakers, actually aren't all about math. And anyway, lobbing more "math" and "science" at these kinds of communication challenges doesn't go very far.

Four Arguments for Planned Phaseout

There are at least four compelling arguments for deliberately phasing out fossil fuels. These arguments are suited to different audiences and interests, and any one of these arguments can make a convincing case for planning phaseout.

The Public Health and Environmental Justice Arguments

The sound of a child gasping for air. Headaches, nausea. Injury to the airways and lungs—inflammation, cell death. Methylmercury-contaminated fish that cause developmental effects in the yet-to-be-born: all part of the cost of fossil fuels. Burning coal produces various pollutants: sulfur dioxide, particulate matter, oxides of nitrogen, mercury, arsenic, chromium, nickel, other heavy metals, acid gases, and more. Extraction and processing of oil and gas also create zones of air and water pollution, as well as toxic waste. The toxicity runs through the whole chain from extraction to disposal. Does that sound too melodramatic? It's probably not dramatic enough. A recent study found that in 2018, air pollution from fossil fuels was responsible for 8.7 million deaths. Pollution accounted for one in five of all deaths that year—about one in ten in the United States and Europe and about a third of the deaths in eastern Asia.[2]

The focus on net-zero emissions occludes these harms, which are experienced disproportionately by people of color and the poor. Workers also bear unique risks. This is a global pattern. Energy scholars Noel Healy and colleagues refer to these harms as hidden "embodied energy injustices," which are rarely accounted for.[3]

The costs can also be measured in dollars, as crude as that is. A report from the Centre for Research on Energy and Clean Air puts the economic and health costs of air pollution from burning fossil fuels at $2.9 trillion in 2018 (from work absences, years of life lost, and premature deaths)—that's $8 billion per day.[4] The researchers quantified the impacts form nitrogen dioxide, ozone, and fine particulate matter, which is the most choking.

The refutation to public health and environmental justice arguments is that with reform, a cleaner version of fossil fuel production is possible. The refutation to *this* refutation is that fossil fuel companies have had plenty of time to clean up their act and have actively resisted it. It's certainly technically possible to produce fossil fuels more cleanly, but what is the evidence for its social likelihood? Fossil fuel companies have lost the credibility to set the terms of their own phaseout. Part of this involves the way fossil fuel companies have been going about exit—by creating spinoffs that go bankrupt and leave taxpayers stuck with cleanup costs. Peabody Coal looks like it may go bankrupt for the second time in five years. Fossil fuel companies are not even cleaning up their own waste right now in the global North, where regulations are relatively strong, which doesn't portend well for cleaning up their mess in the global South.

Take the case of California: its oil industry is in decline, with around 35,000 wells sitting idle. These idle drilling sites can contaminate water and emit toxic fumes and flammable gases. If the companies go out of business, the state will need to find the money to remediate them. Fossil fuel companies are supposed to post bonds to make sure wells are plugged and cleaned up, but a stellar investigation by the *LA Times* and Public Integrity found that California's seven largest drillers only have funds accounting for about $230 per well on average. How much does it actually cost to clean up a well? Between $40,000–$152,000.[5] Companies have given the state $110 million to clean up onshore wells, when in reality it could cost $6 billion. Meanwhile, more than 350,000 Californians live within 600 feet of those old wells and are vulnerable to health impacts.

This abandonment of stagnant wells in itself is another case for managing the phaseout of production, not leaving it up to companies to let production run its desperate course. State regulators at the California Geologic Energy Management Division (CalGEM) see their role as managing decline, but it doesn't seem they actually have the tools to do it. California Resources Corporation, which is one of the state's largest producers, had cleanup costs that are much higher than its total market value. The company was created in 2014 when Occidental Petroleum Corporation spun off its California assets and got rid of environmental liability in the process. The California Resources Corporation went bankrupt in 2020, with $5 billion in debt, and reemerged from bankruptcy three months later. Thousands of deserted and orphaned wells are still marring the landscape.

And Occidental Petroleum? In its third quarter 2020 earnings call, the president detailed the company's low-carbon strategy, which is net-zero emissions associated with operations before 2040, and an ambition for net zero associated with the use of their products by 2050, which is rare for a fossil fuel company. The world's largest direct air capture plant is being funded by Oxy Low Carbon Ventures, a venture capital arm related to the company, along with a venture capital firm. "And in 10 to 15 years, we expect that the cash flow and earnings from a business of this type could be similar or more than what we get from the chemicals business," the president, Vicki Hollub, reported. [6] Those calculations are based on the idea that net-zero targets will get some follow through. As Oxy's senior VP and CFO said:

> And I think underlying it, why we believe it will be commercially viable long term is because in order to achieve the 1.5-degree goal, it can't simply be done through emission reduction. We firmly believe it has to be done through the capture and sequestration or capturing usage of CO_2 to make that goal a reality. And I think what differentiates Oxy's approach to this is we also believe that fossil fuels have a role in the energy portfolio of the world long term. And this is a way to take the carbon footprint of those fossil fuels, keep them part of the portfolio and still generate a low neutral, even negative carbon fossil fuel molecule.

This is a vision for a whole new industry that companies like Occidental are poised to be at the forefront of. The vision may not be durable—activist investor Carl Icahn has battled with the president around the direction of the company. But as President and CEO Vicki Hollub told investors:

> Globally, there's only 40 million metric tons of CO_2 per year that's sequestered or used. And if you look at what the IEA [International Energy Agency] model says about what's going to be needed, what's going to be needed is anywhere from 5.6 billion metric tons to 10.4 billion metric tons. So that's up to—that's going to be more than 250x what we're doing today is going to be needed in the future, and so that's going to make this business incredibly important.[7]

What about those thousands of oil wells that are still leaking toxic fumes into people's homes? This is what Representative Nanette Diaz Barragán, who represents a district in Los Angeles, asked about at a congressional hearing on legislation on carbon capture, utilization, and storage—legislation that makes projects like Occidental's possible. "The climate crisis does require urgent action ... I think there are merits to carbon capture technology for preventing greenhouse gas emissions," Barragán said. But she is concerned about environmental justice, about communities of color in proximity to hazards. "You can see our kids walking around with inhalers around their necks." She's been fighting for the closure of some urban oil wells, not understanding why they are needed in people's backyards. Barragán then asked a key question: "In a carbon constrained future, could carbon capture and storage keep a facility open longer in a disadvantaged community, and would there still be the emissions of pollutants, of sulfur dioxide particulate matter and mercury, into the community?"[8]

It's great that we have a process where representatives can ask these questions. But one of the points I want to make here is that this future is underway. The legislation already being drafted is shaping a future industry of carbon removal and a fossil-fueled path toward net zero through lower carbon fossil fuels. And right now, there is no real plan for dealing with the toxic legacy of fossil fuel extraction. Both an unmanaged phaseout and a net-zero scenario with high production leave environmental justice communities with the mess. This is the current status quo even in one of the world's most environmentally progressive jurisdictions.

The Suffocating Innovation Argument

The second argument for phaseout is that incumbents, such as fossil fuel companies, have a history of blocking new things. They suffocate innovation. Better, newer technologies do not magically win over because of their inner merit—that is an evolutionary delusion about how the world works. The model of innovation here is broken and needs adjustment by the state. Fossil fuel interests are standing in the way of better technology; the better technology has been unable to scale because of the entrenched political power of this industry. Without a planned removal of these blocks, we aren't going to get to better technology. Withdrawing power from the fossil fuel industry will liberate innovation.

"What you see too often in Washington and elsewhere around the country is a system of government that seems incapable of action," the President said. "You see a Congress twisted and pulled in every direction by hundreds of well-financed and powerful special interests ... Often you see paralysis and stagnation and drift. You don't like it, and neither do I. What can we do?"

"Energy will be the immediate test of our ability to unite this nation," the President said, asking for "the most massive peacetime commitment of funds and resources in our nation's history" to develop energy sources. This included the commitment:

> I will soon submit legislation to Congress calling for the creation of this nation's first solar bank which will help us achieve the crucial goal of twenty percent of our energy coming from solar power by the year 2000. These efforts will cost money, a lot of money, and that is why Congress must enact the windfall

> profits tax without delay. It will be money well spent. Unlike the billions of dollars that we ship to foreign countries to pay for foreign oil, these funds will be paid by Americans, to Americans.[9]

Carter was a pretty forgettable US president, and part of his aims for US energy independence called for more coal, but he did have this target of getting 20 percent of the nation's electricity from solar. As his chief domestic policy advisor, Stuart Eizenstat, said, "I'm quite convinced Congress and the American people want a Manhattan-type project on alternative energy development."[10]

So what happened? Where was the innovation, and why are we still waiting for this 20 percent solar? It wasn't a hippie dream, and it wasn't technically impossible. Even two Harvard Business School analysts thought that 20 percent solar by 2020 was a reasonable goal. In the 1979 report *Energy Future*, Robert Stobaugh (who joined Harvard Business after corporate roles in Esso and Monsanto) and Daniel Yergin (fossil fuel historian, currently vice chair at an energy investment firm) wrote that institutional obstacles and market imperfections needed to be eliminated so that government policy could help the markets allow energy conservation and solar to be attractive alternatives to fossil fuels. They explained that the marketplace was distorted. These Harvard Business School guys came to the conclusion that conservation and solar energy were sensible sources and that government directed incentives, like tax incentives and changes in building codes, would be required.[11] They could foresee that innovation would not "naturally" happen.

Popular ecologist and environmentalist Barry Commoner was far more strident in his take on how fossil fuels were stifling innovation. He slammed the National Petroleum Council's 1973 US Energy Outlook for saying that solar couldn't be put into large-scale use in the next fifteen years because it was so diffuse and intermittent, noting that the critique of solar being too expensive was never extended to nuclear fusion, which would be far more expensive than solar.[12] Indeed, the support for these sociotechnical imaginaries was very uneven. A 1973 report by the US Atomic Energy Commission in response to a presidential request, *The Nation's Energy Future*, recommended a five-year, $10 billion research program, of which $200 million (2 percent) would go to research on solar—compared to $1.45 billion on fusion and $2.84 billion on developing a breeder reactor.[13] Yet another government report from that time (Solar Subpanel IX report, chaired by Alfred Eggers of the National Science Foundation) outlined a $1 billion Solar Energy Program aimed at renewables providing 10–30 percent of the nation's energy by 2000 and up to 50 percent by 2020, including wind, biomass, and solar photovoltaic.[14] (In our present darker timeline, 2020 solar and wind provide just around ~4 percent of US energy.)

Solar was a choice half a century ago, and the previous generation did not make this choice. The plan was there. It wasn't even that expensive. But it wasn't adopted.

No decade exemplifies missed opportunities to commercialize renewable energy more than the 1970s. The price of oil tripled, and there were shortages, lines, and rationing in the United States. Reductions in the supply of heating oil led President Richard Nixon to request that people lower thermostats by at least 6 degrees for a national daytime average of 68. "Incidentally, my doctor tells me that in a temperature of 66 to 68 degrees, you are really more healthy than when it is 75 to 78, if that is any comfort," he assured listeners in a 1973 address, adding:

> Now, some of you may wonder whether we are turning back the clock to another age. Gas rationing, oil shortages, reduced speed limits—they all sound like a way of life we left behind with Glenn Miller and the war of the forties. Well, in fact, part

> of our current problem also stems from war—the war in the Middle East. But our deeper energy problems come not from war, but from peace and from abundance. We are running out of energy today because our economy has grown enormously and because in prosperity what were once considered luxuries are now considered necessities ... The average American will consume as much energy in the next 7 days as most other people in the world will consume in an entire year. We have only 6 percent of the world's people in America, but we consume over 30 percent of all the energy in the world.[15]

Given the context of worries about energy supply, not to mention a growing interest in ecology, you would think that in the 1970s, renewable energy developers could have ridden these waves straight toward the goal of 20 percent solar by 2020. The tools were there. Instead, renewable energy stagnated through the 1980s. Why? Fossil fuels blocked the development of what was better. This was true in the 1930s and 1940s, when Henry Ford and others were trying to get ethanol off the ground. It was true in the 1970s, when the Carter administration installed thirty-two solar panels on the roof of the White House, powering the cafeteria and laundry, only for Reagan to take them off and put them into storage. And it's still true today. The intentional obstruction has been well chronicled. Why allow them to keep stifling progress?

The Insurance Argument

Deliberate phasedown can also be viewed as a kind of insurance against the failure of the dominant model of climate policy. This insurance argument is quite elegant, and may appeal to a certain kind of thinker, one who likes to logically assess risks. It is sketched out by Norwegian economist Geir B. Asheim and colleagues in a 2019 comment in *Science*, "The Case for a Supply-side Climate Treaty."[16] Basically, efforts that focus on combustion of fossil fuels are demand-side policies—carbon prices reduce demand for fossil fuels. Phaseout is an approach that addresses the supply of fossil fuels. Asheim and colleagues describe four basic scenarios for the future of fossil fuels when a carbon price, like a tax or a market mechanism, is finally put into place.

1. Companies continue to invest in fossil fuel exploration expecting that carbon price policies will be effective, and they are.
2. Companies continue to invest in fossil fuel exploration expecting that carbon price policies will work, but they don't.
3. Companies expect climate policies to be ineffective, but they actually work.
4. Companies expect climate policies to be ineffective, and as predicted, they aren't.

In the first two scenarios where fossil fuel companies expect carbon prices to work, Asheim and colleagues explain, supply-side phaseout measures are superfluous for restricting investments in fossil fuel exploration. That's because fossil fuel companies are already taking into account future demand-side policies when they make their investments, so the supply-side measures aren't really making a difference. The companies weren't going to develop further sources anyway. This means that the supply-side treaty didn't do any harm. The problem with scenarios one and two is the "Green Paradox": the expectation of the impending price might kick off a "race to extract" reserves. The value of a supply-side policy in these scenarios is to mitigate this Green Paradox.

In the third scenario with pessimistic companies but carbon pricing that works, the supply-side measures aren't needed to avoid dangerous climate change either, since the carbon pricing climate policy is working. The problem here is for the companies: their misguided expectations led to unprofitable investments in fossil fuels. In this scenario, the supply-side plan can actually help fossil fuel companies, because they know what to expect—in a sense, the policy does them a favor in helping them avoid bad investments. It's in the fourth scenario that the supply-side treaty is most needed: demand regulation fails, but the supply-side treaty ensures a safe climate. The supply-side measures are an insurance for a policy based on demand-side carbon price measures. And in the first three scenarios, the supply-side measures are either not harmful to companies—since they align with their plans—or they actually help the fossil fuel companies plan and avoid bad investments.

The Rebalancing Power and Ending Corruption Arguments

The main reasons to end fossil fuels may not even rest on climate change. These arguments for ending them are about the forms of power they enable; the oppression and degradation they are entangled with. Fossil fuels support corrupt and authoritarian regimes. Ending fossil fuels could starve these regimes of fuel.

Corruption, authoritarianism, oppression, conflict—how are all of these entangled with oil and gas? And are fossil fuels always entangled with them, or is there a "good" version possible? Scholars have studied the "resource curse," exploring why countries with petroleum wealth often end up poorer than their counterparts, though this literature finds that the effect is conditioned by governance factors.[17] An academic review found three important effects of petroleum wealth: it tends to strengthen authoritarian regimes, leads to heightened corruption, and helps to trigger violent conflict in low- and middle-income countries, especially when the oil lies in the territory of marginalized ethnic groups.[18] The effects of the petroleum "resource curse" seem to be recent, emerging after the 1970s.

The resource curse framework points out that oil and gas do have some special properties. Oil is a "high rent" industry, meaning that there's a big gap between the cost of producing the good and what it can be sold for. Oil commands such a premium that's it's like winning the lottery with free money. During the oil price boom from 2008 to 2014, as much as $9 trillion in new money flooded into the oil industry.[19] As Alexandra Gillies points out in her book *Crude Intentions: How Oil Corruption Contaminates the World*, this created unprecedented opportunities to make money. Rents are basically like free money that governments can capture. With this "rentier effect," leaders can slash taxes and increase public goods, which reduces dissent and makes autocratic governments more stable. Oil is also volatile, which makes it difficult for governments to manage revenues. All of these dynamics create an environment ripe for corruption.

The reach and depth of corruption Gillies details in *Crude Intentions* is staggering. Some cases are well known: Shell and Eni were charged with bribery in a single deal that allowed political elites in Nigeria to make off with nearly a billion dollars, in a country where corruption keeps the Nigerian people from benefiting from their oil resources. But it's not just Nigeria. Corruption scandals have brought down leaders in Slovakia, South Africa, and Spain. Corruption can be bribes, fund-skimming, tax avoidance. And sometimes, the corruption is legal, Gillies points out. In the United States, the oil industry can essentially bribe politicians through election campaigns. In fact, she illustrates what she calls a "maddening paradox" in several countries: democracy drives corruption, which is motivated by political competition, and then democracy comes to the rescue, checking the corruption.

There's not actually a curse on the commodity, but good and bad management of resources, scholars point out. Political scientist Thea Riofrancos, for example, explains that the resource curse framing is deterministic, and it offers a vision of an ambivalent state and disempowered society. Her study focuses on "resource radicalisms": the ways social movement activists build critiques of extractivism and enact resistance.[20]

These resource radicals can change the fate of extractive projects. This is an important check on the idea that resources necessarily are cursed with corruption, as the determinism inside the "resource curse" idea can obscure the politics and political possibilities. As Riofrancos states, the rentier state and resource curse frameworks can be "elite centric."

Talking abstractly about "oil barons" or "resource curses" or focusing on companies is still somewhat removed from the everyday experience of oppression that people under some resource-reliant regimes suffer. Even though the governments are at fault, the oppression can be linked with oil, and it is also implied in its consumption. It's bad enough that filling up your tank causes climate change, but also political oppression—who wants to support dictators buying mansions in Malibu or Miami Beach when that money should have gone to health care in Equatorial Guinea? Who wants to turn on the gas to make an espresso on their stove and enhance an oligarch, or entrench the power of a guy who just rewrote the rules to keep himself in office until 2036? We talk of "blood diamonds," not so much of "blood gasoline." The entanglement is hard to trace, but that doesn't mean it's not real.

Beyond corruption and repression, there's also conflict, though the relationship between fossil fuels and conflict is complex. Rachel Maddow, in her book *Blowout*, describes how the technological leap in oil and gas production caused by the shale revolution, primarily in the United States, stranded Russia. Maddow traces "the foreign trolls in your Facebook feed" and "the new world disorder" to the geopolitics of energy.[21] Other scholars point out that it is the ideas about oil, not some inherent mojo in the commodity, that lead to this pervasive notion of oil-as-power (and worthy of conflict). Political science professor Robert Vitalis makes similar contentions in his book *Oilcraft: The Myths of Scarcity and Security that Haunt U.S. Energy Policy*. Oilcraft, he writes, is like witchcraft,

> a modern-day form of magical realism on the part of many, diplomats included, about a commodity bought and sold on the New York Mercantile Exchange and elsewhere. The same as copper, coal, rubber, palm oil, tin, and so on, all of which were once imaged as vital, too.[22]

The notion of oilcraft modifies how we might think about this reason to phase out fossil fuels. It's not that they inherently have something about them that leads to conflict; it's that the story the energy intelligentsia tells itself about them leads to conflict. For example, think about the case of oil traveling through the sea lanes in the South China Sea: half the world's oil tanker shipments go through these sea lanes, going to China as well as Japan and South Korea, and they are now a highway for natural gas as well.[23] It's possible that control of these sea lanes could be a flashpoint in conflict, and so an assemblage of professionals prepares for that possibility, perhaps even helping to create it. This is just one of many instances in which ideas about the control and movement of fossil fuels play into geopolitical instability.

Focusing on net zero and on climate-related reasons to phase out fossil fuels misses all this: they do actually have a role in oppression, corruption, poverty, and perhaps also war. Without being deterministic about it—and saying that some evil energy inheres in the commodity—we can still appreciate that oil and gas have developed a particular social context that entangles them in oppression, corruption and conflict. All this is arguably a much bigger, or at least more immediate, threat than climate change. It exists right now and impacts the daily lives of millions, if not billions, of people. There's a nonzero risk of conflict over the endgame of fossil fuels, which seems on par if not worse, in terms of human suffering, than climate change. Why do most arguments for ending fossil fuels focus on emissions and gloss over these dynamics? They seem a far more certain and obvious reason to phase out this industry in a planned way.

Here, we have several decent arguments for deliberately winding down the fossil fuel industry. They beg the question: If phaseout is so obviously necessary, why haven't we been talking more about it? Certainly, the "Keep It In the Ground" movement, environmental justice movements, and Indigenous climate advocates have, but the mainstream conversation still isn't commensurate to the challenge. Reasons include the obvious power of the fossil fuel industry and activist preference for folk politics and localism or ground-up approaches rather than top-down planning of phaseout. Phaseout is also not a mediagenic topic: it's hard to capture detailed demands about specific ways to phase fossil fuels out in a clip or a picture. The sound-bite version of phaseout sounds naive.

At present, despite the evolution of talk about supply-side measures and winding down coal, talk about phaseout tends to be either thin and technocratic or simplistic and sloganlike. But there are plenty of compelling reasons to consciously end the production of fossil fuels. What we've covered so far isn't even all of them. You could think of additional reasons: We need to phase out fossil fuels because failing to do so means that oil companies will pivot to petrochemicals, and the world will be awash in plastic. Or: We need to phase out fossil fuels to mitigate the risk of sudden destabilizations of the climate system. Or: We need to plan a phaseout of fossil fuels because otherwise the financial system might be destabilized by their volatility in a decarbonizing world. Even capitalists will find reasons to do this if they look.

The Risks of Phasing out Fossil Fuels

At the same time, deliberately phasing out fossil fuels does come with serious risks. The argument here is not that these risks are negligible. They are real. It's just that the risks of *not* planning the end of fossil fuels are worse. We have to catalog these risks and figure out ways of mitigating them.

First, there is the risk of communities not having enough energy. This is a risk that exists anyway in the energy transition (it is, in fact, the status quo in many places). It is possible that a botched phaseout could make things worse. How much worse? We don't know. It needs to be studied. In one manifestation of this risk, there is simply not enough energy produced, because of underinvestment in fossil fuel production coinciding with an undersupply of renewables. In another manifestation, fossil energy is being produced but it is inaccessible to people because it has been made too expensive by phaseout policies. If fossil fuel companies are forced to limit production, naturally they would want to compensate for these losses by raising prices—thus there would be pressure on the state to subsidize fossil fuel consumption for people who still need fossil fuels for getting to work or heating their homes. (In fact, resource exporters might prefer supply-side policies: they could in practice behave like OPEC, leaving parts of fossil fuels in the ground and raising prices on the rest.[24])

These risks of inaccessible or expensive energy carry follow-on effects, in terms of creating unrest and state stability. They could also spark domestic backlash and erode support for climate action, not to mention other progressive policies. In the 1970s, New Deal tools of government intervention to solve crises, like price controls, rationing, and antitrust actions, were normal. Conservatives were able to credibly argue that it was the regulations themselves that were stifling US energy self-sufficiency. The energy crisis of the 1970s ultimately helped shift US politics to the right. Historian Megan Jacobs argues that the failure to address the energy crisis, with the sense of vulnerability it created, contributed to the erosion of Americans' faith in government, illustrating the limitations of government power.[25] Measures that make energy less affordable can also provoke popular opposition. Gas tax increases were a driver in the spontaneous grassroots mobilization in the French *gilets jaunes* (Yellow Vest) movement of

2018, in which people wearing yellow vests occupied traffic roundabouts. The movement drew people from both left and right living across fragmented social spaces at the edges of urban centers and suburbs, points out Stefan Kipfer—people whose lives have been shaped by automobilization and decentralization. Their protest of the tax hikes weren't necessarily about being anti-ecological, but about refusing fiscally regressive market measures to confront climate change.[26] Doing phaseout in a regressive way could set back climate action.

A botched attempt at managed decline could backfire not just in terms of politics, but in terms of emissions. It could kick off the "race to extract," what Asheim and colleagues call the "Green Paradox," or what other scholars have called a "panic and pump" response from fossil fuel companies.[27] These risks highlight how important it will be to plan well. In a world full of risks, these risks seem easier to mitigate and cope with, compared to the alternative risks in ignoring managed decline and getting an unmanaged decline or no decline at all.

Looking at phaseout with several different lenses can bring dimension to it. Phaseout is not just one problem—it's several kinds of problems at the same time. We have to be able to see and address all these dimensions, because treating it through one lens will fail.

What follows is a framework for looking at phaseout. For a long time, I didn't know what "framework" was even supposed to mean. It still sounds like jargon to me. But here I mean it as a tool for seeing. Imagine five frames, each of which helps you notice something different about a mysterious object. The frames draw our attention to elements that we might not otherwise spend time on. They allow us to consider each dimension with full attention: dimensions of culture, infrastructure, geopolitics, code, and political power.

Part II. Five Ways of Looking at Fossil Fuel Phaseout

6

Culture

The highway stretching out into the distance: freedom, expansion, limitless growth. There's a way in which managed decline sounds like failure, defeat. "Petromelancholia" is a term coined by Stephanie LeMenager: "loving oil to the extent that we have done in the twentieth century sets up the conditions of grief as conventional oil resources dwindle."[1]

Phaseout is a problem of cultural change: it means evolving new values, beliefs, practices, and rituals. "Fossil fuels will make a moral transition in parallel to the material transition," write Thomas Princen and colleagues in their 2015 book *Ending the Fossil Fuel Era*: "Much as slavery went from universal institution to universal abomination and as tobacco went from medicinal and cool to lethal and disgusting, the delegitimization of fossil fuels will flap the valence of these otherwise wondrous, free-for-the-taking hydrocarbons."[2] There are many cultures to change, Princen observes: the organizational cultures of fossil fuel companies; the industry culture; the high-finance culture of economic policy; the economic culture of growth; the consumer culture.

Too often, the focus is limited to change in consumer culture. Discussions about "behavioral change" constrict the terrain of thinking about cultural change. Matt Huber's *Lifeblood: Oil, Freedom, and the Forces of Capital* offers a more useful view of oil as a social relation. Huber explains the real subsumption of life, in which living labor appears as capital itself. One's own life is seen as "an individualized production of hard work, investment, competitive tenacity, and entrepreneurial 'life choices,'" expressed through things like a home, a car, and a family, and specific regimes of energy are the product of these struggles. "The biggest barrier to energy change is not technical but the cultural and political structures of feeling that have been produced through regimes of energy consumption," Huber observes. Fossil fuels produced an unprecedented type of individuated power and control over everyday life, which made people feel free and on their own. Yet the cars, homes, roads, and plastics around us, he writes, are products of broader social relations.[3]

Petro-nostalgia is also a thing. Cara New Daggett's work chronicles the petro-nostalgia of Donald Trump and new authoritarian movements. Misogyny and climate denial are often regarded as different dimensions of new authoritarian movements, she observes, but "a focus on petro-masculinity shows them to be mutually constituted, with gender anxiety slithering alongside climate anxiety, and misogynist violence sometimes exploding as fossil violence."[4] How to change these patterns? Daggett calls for putting postcarbon movements into conversation with the postwork political tradition. Without this fossil fuels = jobs equation, a lot of the arguments for continued production fall apart. Aligning with postwork movements could help environmentalists produce an alternative political vision of pleasure, she argues. There is a clear need for some attractive cultural vision and set of values to replace petroleum-fueled consumption.

Change can also be encoded in language. Princen chronicles the lexicon of fossil fuels—rush, boom, frenzy, gusher, jackpot, man camp, wildcatter, roughneck—words that connote excitement and daring but are short-term, pointing out that these contrast with terms like permanence, attachment, blessing, gratitude, stewardship, or common good.[5] There are all kinds of needs for cultural change to equip us for phaseout; here I'll point out just two: increasing comfort with endings and making planning into a wider cultural value.

A Vocabulary for Endings

Loosening ourselves and each other from the "structures of feeling" that have been produced through particular regimes of energy consumption is hard enough. But confronting endings is an even more widespread cultural problem. Our culture sucks at ending things.

The amount of things that need to be ended is mounting. Not only do we need to end things that are harmful, like fossil fuels or pesticides or plastic bags; we need to end things that are doomed. The latter demands a shared understanding of what's doomed. When it comes to climate change, there are parallels between managed decline and managed retreat. It won't just be moving away from coastlines or wildfire-prone areas—it will be moving from areas where aquifers are depleted, ski towns where it no longer snows. We need a wider cultural shift that will enable us to manage all of this change and these various endings—not something that dismisses them as "all okay," indicating that we should simply "Be resilient!" either, but something that allows for grief as part of the process.

There's a dangerous feedback loop: as we lose more and more, and as our quality of life slips—life expectancy, closed businesses, degraded infrastructure—the more we're afraid of slipping and cling to conservative ideology ("make America great again"). For some nations, like the United States, imperial decline is all mixed up in the decline of fossil fuels.

There's also a more liberal version of this grasping-in-the-face-of-decline that clings to innovation as the remedy. "What passes for innovation is actually innovation-speak," caution Lee Vinsel and Andrew R. Russell in their book *The Innovation Delusion*. They call innovation-speak "a dialect of perpetual worry."[6]

> While it is often cast in terms of optimism, talking of opportunity and creativity and a boundless future, it is in fact the

> rhetoric of fear. It plays on our worry that we will be left behind: Our nation will not be able to compete in the global economy; our businesses will be disrupted; our children will fail to find good jobs because they don't know how to code.[7]

Notably, the rate of innovation has decreased since around 1970. Vinsel and Russell draw on economist Robert J. Gordon, who observes that technological improvements since the 1970s have been in computing, cellphones, and digital platforms. Compared to the 1870–1970 period, with advances like electricity, sanitation, pharma, plastics, concrete and steel, airplanes, and computers, there's the question of whether startups of today are just playing out technologies created before the 1970s rather than creating genuinely new ones. Innovation stalled as innovation-speak rose, they observed.

While it might be reflexive to reach for innovation when grappling with decline, we may need to innovate but simultaneously move in another direction. "The innovation mindset has led to a devaluation of maintenance and care, with disastrous results," write Vinsel and Russell.[8] Maintenance, they posit, is in some ways the opposite of innovation. By maintenance, they mean the overlooked and underpaid work of keeping daily life going, caring for people and things, sustaining what we've collectively inherited. Care and maintenance are alternative vocabularies for talking about what we choose to maintain and what we choose to retire. While maintenance on some level is still about keeping things going, fixing our decaying infrastructure relies on the same kinds of capacities to plan that we'll need for managing the decline of unwanted industries and practices.

A Culture That Plans

Planning isn't just an institutional or financial capacity; it's a cultural one. How does a culture learn to plan? Here the inquiry isn't just about how planning becomes formalized, a disciplinary culture, but how people in the broader culture think about the role of planning in their lives.

Planning may have been birthed in war; let's hope that there's another way to develop it. Economic planning found expression in Germany during World War I (*Planwirtschaft*) and was followed consciously by the Russian Bolsheviks after 1917, who launched a grand experiment in economic planning.[9] This is described in Leigh Phillips and Michal Rozworski's book *The People's Republic of Walmart: How the World's Biggest Corporations Are Laying the Foundation for Socialism*, and also discussed in Andreas Malm's *Corona, Climate, Chronic Emergency*.[10] As Phillips and Rozworski recount, Lenin and other Bolsheviks hadn't thought much about how the economy would be run after they came to power.

> Centralized planning arrived in drips and drabs, on an ad hoc basis—often in reaction to the disruption or collapse of normal market relations and acute shortage as civil war spread throughout the country—rather than through the stepwise rollout of a comprehensive strategy for replacing the market.[11]

The winter of 1917–18 was brutal: workers left the city to seek food, and factories closed. Supplies of consumer goods, raw materials, and fuel ran out. This drove the creation of Vesenkha, the Supreme Council of the National Economy, and later Gosplan, which was established in 1921 and charged with economic planning, including the first system of national accounts. This was a real innovation. Phillips and Rozworksi explain that the logistical, accounting, and planning techniques that the Soviets developed were adopted by capitalist corporations and are still used today, making a five-year plan not just an operational plan, but a strategic one.

It was only after Soviet five-year planning looked successful that British thinkers became interested in it. In 1932, John

Maynard Keynes hailed state planning as "a new conception ... something for which we had no accustomed English word even five years ago."[12] Planning was so popular that it took on a variety of meanings. Socialist planning in the 1930s and 1940s featured a broader idea of planning, which included health, housing, social security, town planning, and economic planning.[13] This began to change with the scientific, rational planning that arose in the post–World War II era. The rational comprehensive model of planning (initially developed in Germany and introduced in the United States) involved land use and transportation modeling, helping to develop planning as a profession. What evolved was a fragmented profession of city and regional planning, a mashup of land use, housing, transport, environmental planning, and development planning. Economic planning was no longer part of the plan.

Neoliberalism brought a turn away from the role of government and away from regulation. The emphasis during the 1980s was to liberalize planned economies and enact reforms, across Eastern Europe and the former Soviet Union—remove government controls, move to a market economy: the age of *perestroika* and *glasnost* on the evening news. But it wasn't solely neoliberalism that was responsible for this turn away from planning: the counterculture helped. As planning professor Bishwapriya Sanyal writes, these golden years of planning lasted for nearly two decades, placing 1968 as a turning point: "what came under attack were not only the results of planning but also the culture of planning practice ... Planning was now considered too technocratic, elitist, centralized, bureaucratic, pseudoscientific, hegemonic and so on."[14]

The way professionals think about economic planning is a legacy of this: the role of (some) planning involves urban infrastructure and making cityscapes inviting for capital, generally speaking. The rest can be left up to the market. In the broader culture, ideas about freedom are at the heart of the conflict over planning. Friedrich Hayek argued that a market economy was better than a centrally planned one because it could make use of the knowledge of individuals; economic planning would obviate the skills of individuals and lead to totalitarian domination. More than that, though, a central plan would impose the values of the bureaucrats upon individuals. This is still a core fear that people have around planning: the loss of individuality, the imposition of someone else's values. But democratic planning is possible—and essential to the process, say Phillips and Rozworksi.

> A non-market economy is not a question of unaccountable central planners, or equally unaccountable programmers or their algorithms making the decisions for the rest of us. Without democratic input from consumers and producers, the daily experience of the millions of living participants in the economy, planning cannot work.[15]

This will require fundamental transformation of the relations or structures of society—it's not a simple technocratic reform.[16]

The last iteration of Big Planning wasn't any better than Big Science, in terms of its structural flaws, and so it makes some sense to have turned away from it—it was racist and unequal. We have to reinvent planning and also reinvent science, so that they are informed by diverse experiences and needs and work for people, not capital.

This reinvention is poised to happen. As urban planning professor Kian Goh describes, the neoclassical turn in planning that was prevalent from the 1970s through the 1990s gave way to a communicative planning regime that acknowledged communities and power structures and paid attention to variable environmental outcomes. The social transformation tended to be spatially neutral and small-scale. Goh suggests that this may be replaced by a Green New Deal planning regime, which involves state public spending with a focus on structures of power and attention to the previously marginalized:

"A bigger-than-local framework of justice turned toward climate change challenges would compel planners to think about justice as interconnected across political and ecological boundaries, working across often-unaligned administrative areas and watersheds."[17] Meanwhile, ideas are flourishing about how to make science more inclusive and in the service of public aims. From the field of design, there are increasing currents of design justice. Yet economic or industrial planning is still a world apart from, say, landscape architecture. There is a learned separation between these types of planning. But actually, net zero as a goal implies being able to plan in all these areas.

Confronting climate change will require planning everything better than we currently do, from food systems to waste management to energy to land use. For those capacities to grow, they need to become part of our culture. How does that happen? Through education, media, art, making planning into a cool career, mainstreaming it into childhood—all the "executive functioning" stuff that rewards planning in one's own life. Through celebration of planners and plans well achieved, giving celebrity status to people who design and see through equitable plans, activities like community gardening that have an experiential element of planning, field trips to sites of planning ... and so on. We have to build institutions of democratic planning so that everyone can participate in it—or so that we can delegate making the plan to people we choose, not McKinsey consultants and black-box platforms.

What we get from looking through the culture lens: empathy and a flowering of technics for transformation before and beyond policy.

7

Infrastructure

The Somerset coal plant looms over stubby corn and skeletal vineyards. Little lights blink on its smokestacks in the winter fog. Settled on the shore of Lake Ontario, Somerset was the last operating coal plant in New York State, producing 675 megawatts (MW) of power. The plant ceased operations in 2020.

"We're all happy to worry about we are going to shut down coal plants, or we're going to end fossil fuel as we know it. That is a bunch of shit. You can't, unless you deal with the fallout. The social impasse of shutting down fossil fuel is totally BS, it can't be done," argues Paul Schnell.[1] Paul is a naturalist and conservationist who worked at the plant for thirty years, in the material handling department, unloading coal trains and limestone trains. Working in material handling gave him opportunity to observe birds on the 1,600 acre property. Paul is a master raptor bander who specializes in American kestrels. He spent years traveling and giving talks with a bald eagle named Liberty, and he loves to talk about this. He argues, "You give these children an opportunity to see and experience. And this is how you make proselytes for conservation." He's pro-wilderness and vehemently against overpopulation, suburban sprawl, and wind turbines, which he calls bird slicers and bat slicers. "Talk to the people who have done the bird and bat counts around wind turbines. From a half a mile away, a wind turbine looks like a big, clumsy pinwheel, doesn't it? Get close to it. Those blades are ferociously turning at 150 to 200 miles per hour at the tips," he says, pointing out that birds haven't evolved to swoop and swerve around those blades.

Paul's not alone in his ire for the wind turbines. Many houses scattered around Somerset boast anti-turbine signs, demanding "Local control," declaring "Too big, too close." A proposed 200 MW project by a Virginia company would involve forty-seven turbines of up to 591-feet high. While it would bolster the tax base in the wake of the coal plant's closure, some of the local towns have passed laws against the turbines, though state agencies can override the laws. This area also has significant local opposition to large-scale solar, with proposals including one for a 900-acre, 100 MW facility. (As a reference, 900 acres is roughly 1.4 square miles, or 3.64 square kilometers.)

New York State has ambitious climate legislation, including a target of net zero by 2050—and their target is interesting in that it specifies that 85 percent of reductions have to come from emissions, meaning that offsets for residual emissions could be just 15 percent. The state also banned fracking. So it is on its way to figuring out where this clean energy will come from. To compensate for the energy from the coal plant you'd need something like 150 turbines or nearly seven of those solar farms. That would be enough energy to power 105,000–140,000 single family homes, roughly speaking. This county, Niagara, has 88,519 households (many of which are in single family homes), so you can figure this amount could power this county and much of the next. Back-of-the-envelope, the area would be energy self-sufficient if it accepted these plants and built a few more of similar size on top of them. But then keep in mind: urban areas don't necessarily have this build-out capacity and will need to be sourcing renewables from rural areas like this one; in addition, electricity will need to increase to make room for electrifying everything. So it really does add up to noticeable impacts on the landscape, for humans and potentially for non-humans—several times the amount of infrastructure that the community is attempting to reject. A key report done under the Obama administration, *Pathways to Deep Decarbonization in the United States*, found that electricity generation would need to double by 2050 while its carbon intensity is reduced to 3–10 percent of its current level.[2] Moreover, in a high renewables scenario (that is, without relying on nuclear or carbon capture and sequestration), 2500 GW of wind and solar capacity would be required—in 2020, non-hydropower renewables capacity was at 181 GW.[3] While wind and solar are growing, to get from 181 GW to 2500 GW is still a reach. These sound fine as abstract big numbers; trickier when you have to address them at the scale of the land you live and work in.

I used to write off these oppositions as aesthetic—living next to a few spinning monoliths is a pretty small price to pay for a livable climate. But what if the choice isn't between turbines and climate breakdown, but between turbines and cleaner fossil fuels? How many will go for the turbines? Because that's how people like Paul see it. In his view, the coal plant was already more environmentally friendly than this impending wind and solar infrastructure. "When that thing came online, it was a billion dollars, because it was the first plant in New York to be equipped with a dry scrubber and a wet scrubber. The dry scrubber is the electrostatic precipitator. That pulls the solid particles out of the gas path. Then there is the wet scrubber, which is a limestone slurry." The wet scrubber is for removing the sulfur dioxide. "We all had a job to do, and we all realized that it was in everybody's best interest, especially society's, to make the cleanest possible energy with the least amount of pollution," Paul recalls. "The one thing a lot of us had was an environmental ethic. We were raised by parents that gave a damn. And you gave a damn and all these people, for the most part, gave a damn ... You come and talk to me when you spend thirty years in a facility that is always in the top ten, always, and people take pride in that." For Paul, the environmental elite is full of hypocrisy. "I tell you what, their little suit-tie affairs with their expensive big fancy cars, the carbon footprints are double, triple, quadruple, quintuple the rest of the people." But in Paul's view, environmentalists are out of touch with the environmental realities of renewables versus cleaner coal: they're not looking at the full picture.

From this standpoint, why not keep the coal plant running, but balance its CO_2 emissions with carbon sequestration? Would that be more environmentally friendly than renewables? Paul points out that New York is 63 percent forested, thanks in part to the vast Adirondacks, and has millions of tons of carbon sequestration capacity in its forests. You can see how this version of net zero might sound like a compelling plan to people.

Retire, decommission, mothball: the vocabulary of ending old infrastructure, or even newer infrastructure, paints the work in colors of steel and rust. Perhaps "quiet" or the neutral "dismantle" would be better words for the task. Either way, there's a physical aspect to taking things apart (or leaving them to rust, as the case more often is here in the Rust Belt). Then there's the work of building the new, exciting, and daunting at the same time. The trick is to do both this building and unbuilding in a coordinated dance that continues to supply energy and employment and avoids a broader financial crisis. This dance is perhaps the first or most dominant way of thinking about the challenge of just transition. It is the easiest to study, since infrastructure is physical. It is also something that can be understood and planned for using models.

From an infrastructural point of view, the problem is fairly well understood. As the International Energy Agency states in its 2020 *World Energy Outlook*, "Avoiding new emissions is not enough: if nothing is done about emissions from existing infrastructure, climate goals are surely out of reach." If today's energy infrastructure continues to operate as it did in the past, it would lock in 1.65°C.[4] That infrastructure comprises power plants, industrial facilities, buildings, and vehicles—all of that together would still be emitting around 10 gigatons of CO_2 in 2050, even if all the new facilities were generating from cleaner sources. So it's clear that some of the existing stuff has to be put out of service "early," before the end of its life span. These assets then become "stranded." *Stranded*—it sounds like a lost puppy, but it's generally defined as "assets [that] suffer from unanticipated or premature write-offs, downward revaluations or are converted to liabilities."[5] *Suffer*—that's how we tend to talk about these, though it would also be fair to say that they were mistakes that shouldn't have been built, given that many projects were developed when climate change was already acknowledged as an issue.

Stranded resources are resources that can't be used—the "unburnable carbon"—but increasingly, this is also used to describe things like water, metals, or forests.[6] Climate change is quickly becoming a stranding force—both climate change itself, and policies introduced to deal with climate change, can be agents that strand. For example, with the introduction of forest protection policies, forests are transformed into stranded resources that can't be developed for some other use. Researchers Kyra Bos and Joyeeta Gupta suggest that the UN's mechanism of Reducing Emissions from Deforestation and Forest Degradation (REDD+), which compensates forest holders, can be viewed as a mechanism for dealing with stranded assets and compensating for the opportunity costs of not developing a resource.[7] Here, we'll focus on fossil fuels, though it's interesting to keep in mind this broader context and discourse of stranding, because these other changes will influence the way we think about phasing out fossil fuels—managed retreat from coastal properties bears some resemblance to managed decline. The change from using vast areas of land to feed cattle to restoring this land will be a transition that requires as much planning as phasing out fossil fuels does.

Let's look at the different assets that need to be retired, starting with what's involved with the fossil fuel sector. Basically, the sector has three parts. There are upstream activities, which include exploration and extraction—fossil fuel reserves, or stuff underground, are considered upstream assets. The estimated value of stranded assets in upstream oil and gas is US$3–7 trillion.[8] Then there's transport, which includes pipelines, rail lines, ships, liquefied natural gas terminals, and so on; asset stranding here involves canceled projects or projects that aren't fully used. Finally, downstream operations include oil refining, distribution (such as gas stations), and power plants. When it comes to downstream operations, many companies have already divested in retail, but refiners are a big source of stranded costs. Globally, the total value of stranded assets in the power sector alone could be as much as US$1.8 trillion.[9]

So what gets retired first? Coal plants are the most obvious thing, and this is well underway. There are 7,604 coal-fired power plants operating, under construction, or proposed around the world according to Global Energy Monitor's Coal Plant Tracker as of December 2020. It will be comparatively easier to retire these in the United States and Europe because facilities there are older. According to an analysis by civil engineering professor and environmental sociologist Emily Grubert, a deadline for decarbonizing energy by 2035 in the United States—which is what has been proposed by Joe Biden and the Democratic platform—would only strand about 15 percent of fossil capacity-years and 20 percent of job-years, partly because 73 percent of US fossil-fueled generation capacity reaches the end of its typical life span by 2035. The remaining 27 percent would have to close earlier.[10]

The numbers are knowable and countable: Grubert reports that in the United States in 2018, there were 10,435 fossil fuel generators providing 63 percent of US electricity, emitting 1.9 billion tons of CO_2, 1.3 million tons of nitrogen oxides, and 1.4 million tons of sulfur dioxide, consuming 3.2 billion cubic meters of water, operating in 1248 of 3141 counties, and employing about 157,000 people. While this is a lot to phase out, it's possible to make a plan for this. As Grubert highlights, "We have seen before what happens without adequate planning and preparation, such as with the collapse of the U.S. steel industry in the 1970s and 1980s."[11] This context of de-industrialization provides an eminently understandable rationale for planning the transition, not just letting it unfold.

Because there have been so many plants retired already, there's plenty of experience to draw upon in shutting them down. In his book *Power After Carbon: Building a Clean, Resilient Grid*, Peter Fox-Penner explains the tools that help with plant retirement in the US context. Basically, fossil fuel plants can be owned by regulated investor-owned utilities (IOUs) or by independent power producers (IPPs). Early closures of IOU-owned plants are a bit easier, because the utility has other assets and can invest in new replacement power sources; IPP-owned plants are more difficult because the owners are unregulated, for-profit companies, and so they tend to close less often.[12] (This is another example of neoliberal deregulation making this problem harder than it should be.) If a facility is owned by regulated IOUs, several groups are involved in closure: utility stockholders and bondholders (represented by utility management), customers (represented by their regulator and sometimes a consumer advocate), workers and unions, and community leaders. All these stakeholders need to agree to a retirement plan to avoid delays and litigation; this plan involves stuff like financing the replacement power, paying off bonds, providing acceptable profits to equity holders, creating a transition plan for workers, keeping rates reasonable, and compensating the general community.

Fox-Penner describes some of the financial tools that are used, such as creating smaller "regulatory assets" that continue to act like a profit-generating plant after closure, or issuing special securitized bonds that pay off the stream of projected earnings. Good sources for learning more about the existing practices and tools at hand include *Power After Carbon*; the work of the Coal Transitions international research hub; and the Rocky Mountain Institute's work, including *How to Retire*

Early: Making Accelerated Coal Phaseout Feasible and Just.[13]

There's an underlying question here: Who should pay for a mistake? At some point that mistake is intentional, a bad decision made with full knowledge. According to capitalist logics where everything is a competition and the most meritorious companies succeed, the companies that made bad decisions should simply fail. We don't protect companies from the bad decisions made in other domains (except for, apparently, banking)—why should fossil fuels be special? Whichever way you fall on this question, it seems clear that the response should be focused on protecting people from the bad decisions made by corporate leaders. And if we're going to be stuck paying to bail them out later on, we might as well do it sooner, before more climate harm is done.

Stranding assets and resources could be particularly tough for the developing world. Modeling analysis that looks for the most economically efficient outcome finds that Africa would need to leave 26 percent, 34 percent, and 90 percent of gas, oil, and coal reserves respectively untouched; China and India, 25 percent, 53 percent, and 77 percent; and the rest of Asia, minus the Middle East, 12 percent, 22 percent, and 60 percent.[14] Countries like Kenya and Mozambique have recently discovered fossil fuel resources and are planning to use these resources to increase wealth—but they now run the risk that new assets created to extract, refine, and distribute them will be stranded.

The infrastructural lens on phaseout is important, because it calls attention to the spatiality of phaseout and the human geographies of workers and economies that depend on these infrastructures in particular places. But it's also a kind of odd lens, because one would expect it to be clinical. Yet the language of asset stranding actually tends to be emotional, placing the listener in a position to identify with the infrastructure. Can you really identify with a coal plant? No, but the discourse often invites you to, centering the plant and its owners as objects of sympathy rather than the people who have lost loved ones due to air pollution. It also tends to make an exception out of fossil fuels, eliding the fact that infrastructure and facilities are closed or abandoned pretty much all the time. At the same time, trying to change the tone from funereal to celebratory, or from funereal to clinical, would not be very compassionate to the communities that depend on these plants for tax revenue and jobs.

What we get from looking through the infrastructural lens: many precedents to draw upon in terms of closing fossil fuel facilities and a spatial, place-based understanding of the problem.

8

Geopolitics

Some countries are relatively well set up for an energy transition. Others are not. Debate about the energy transition has mainly taken place in the world's financial centers. However, over half the world's least developed countries have plans to expand fossil fuel production as a lever for economic development.[1] Most of these reserves require the technology and finance of international oil companies to extract—so these developing "stranded nations" have the largest proportion of assets exposed to stranding.[2] What does the world look like if they are unable to phase out fossil fuels?

Imagine: In the green half of the world, there are self-congratulatory speeches about reaching net zero. There is still some amount of fossil fuels consumed, but there are also vast direct air capture plants spinning to compensate for this amount, and wide swaths of land have been reforested to generate land-based removals. Meanwhile, in the fossilized half of the world, energy-reliant economies are still cranking out fossil fuels and selling them to nations that have not been able to organize or afford vast renewable infrastructures and the grid needed to make them work.

Given the abysmal record of the global North when it comes to technology transfer and climate finance, this is a fairly plausible scenario. As World Bank analysts point out, "climate action in net fuel importing countries could lead to so-called 'dirty' diversification with the relocation of emission-intensive industries to [fossil fuel developing countries]."[3] A move away from fossil fuels could lead to increased production in another area, either through a "panic and pump" or "race to extract" before decarbonization deadline strategy by exporters,[4] or a "carbon dumping" dynamic, where carbon intensive industries go to countries with laxer environmental policies. These scenarios have plenty of domestic implications for people—and the climate—but they will also shape how countries interact with one another.

To understand the geopolitics of winding down fossil fuels, we must understand: Who is reliant on oil? Who has the ability to switch to another source of energy? Who has the ability to change their economy to another basis?

With this information, we can then ask: What will fossil fuel reliant countries do in response to the demands of phaseout, or to phaseouts in other countries, given how they are governed? Who will help them, and how? Who will use the new configuration of energy against them? And what do those answers mean for strategic thinking about encouraging phaseout?

Vulnerability to Phaseout

Many countries are dependent upon oil and gas. But some are actively working toward economic diversification. Others are more vulnerable or less prepared. According to a World Bank analysis, Iraq, Libya, Venezuela, Equatorial Guinea, Nigeria, Iran, Guyana, Algeria, Azerbaijan, and Kazakhstan are the least set up for a low-carbon transition.[5] The vulnerability is most acute with small oil and gas producers in sub-Saharan Africa, North Africa, Latin America, and the Middle East that have "not yet diversified their exports or undertaken a structural transition towards knowledge-intensive, low-carbon economic growth," in World Bank-speak. They also point out that many of these countries are also extremely vulnerable to climate change. Saudi Arabia, Venezuela, Brazil, and Nigeria are also viewed as geopolitical losers in phaseout; Europe and Japan are potential winners.[6] Indeed, some analysts point out that both the United States and Europe would have strong positions in a post–fossil world.[7]

The fact that the United States has become an energy superpower, and could benefit from ending some fossil fuels but expanding cleaner gas production, is important context for making the demand for ending fossil fuels. In the early part of the twenty-first century, two things have roiled the geopolitics of energy. One is the shift to renewables, which is a familiar story. The second is the recent revolution in shale. In the 2000s, technological innovations in horizontal drilling merged with innovations in hydraulic fracturing, or injecting sand, water, and chemicals under high pressure to crack rock. Vast quantities of new gas and oil became available. Coal was priced out. Liquefied natural gas become a global industry. Since 2000, liquefied natural gas (LNG) demand has quadrupled, the number of countries exporting LNG increased from 12 to 20, and the number of importing countries rose from 11 to 40.[8] The shale revolution also rocked prices. The oil price had been stable at just over US$100 a barrel from 2011 to 2013. But at the end of 2014, the oil price dropped. Low prices showcased the vulnerability of petrostates, many of whose budgets had been calibrated around $100 per barrel. Revenues from oil exports fell from $321 billion in 2013 to $136 billion in 2016. People expected OPEC to cut production, but only Saudi Arabia had the capacity to, and it didn't want to help Iran or lose market share and money. Mexico and Russia refused to cut production. There was a glut of oil, with prices dropping below $30 a barrel. Oil exporters suffered; countries like Russia and Saudi Arabia began draining funds.

The United States, on the other hand, benefitted tremendously from the shale revolution. From about 2009, US production of oil and gas increased dramatically. The shale revolution lifted up the US economy—between 2009 and 2019, the increases in oil and gas accounted for 40 percent of the cumulative growth in US industrial production, supporting millions of jobs across the country. Government and state revenues from this boom are estimated at $1.6 trillion from 2012 to 2025.[9] The United States went from anticipating needing to import gas to being an exporter of gas, and all of this improved its position in the world economy. So the United States would presumably have a strong position in a decarbonizing world, especially if gas is treated as a bridge fuel. But how do we really know how phaseout would affect different countries?

It's actually a bit tricky to track how vulnerable a country is to phaseout. It's hard to find a clear-cut indicator to measure resource dependency, observe energy analysts Dawud Ansari and Franziska Holz. Typically, people measure the share of natural resource rents as a percentage of GDP—if you just looked at that, you'd see that Kuwait and Iraq range above 30 percent, but Bahrain and Colombia have just 3 percent, Nigeria 5 percent, and the United Arab Emirates 11 percent. However, you have to look at the shares of fuels in exports, too. Colombia's coal exports make up more than 50 percent of merchandise exports; oil accounts for 96 percent of Nigeria's. Those are critical sources of foreign exchange. Then look at the share of government budgets supported by resource revenues: for the United Arab Emirates and Bahrain, more than 2/3 of their budget is derived from resources—for Saudi Arabia it's over 90 percent.[10] There are a lot of ways that countries are locked into fossil fuels.

But some powerful leaders see the need for eventual change: as the Crown Prince of Abu Dhabi and de facto leader of the United Arab Emirates, Mohammed bin Zayed, said, "In 50 years, when we might have the last barrel of oil, when it is shipped abroad, will we be sad? If we are investing today in the right sectors, I can tell you we will celebrate."[11] The timeline for this diversification is notable. The CEO of the state oil company, the Abu Dhabi National Oil Company, has said that low carbon oil will play a central role in the energy transition, noting that the United Arab Emirates is positioned to provide lower carbon oil as long as the world still uses hydrocarbons, and that it is also exploring CCUS and hydrogen.[12]

However, there are not many examples of countries diversifying. While Malaysia and Indonesia diversified as manufacturers, and Dubai attracted foreign investment in services and business, these are not replicable models across the board and come with caveats.[13] Malaysia and Indonesia never produced as much fossil fuels as other nations, and the Dubai model is not a great one to replicate, given that it depends on racialized and deeply underpaid immigrant labor.

Saudi Arabia is an archetypical petrostate[14] and another key example of a country attempting to diversify. It has grown rapidly since 1990, with a population doubling to 33 million since then, and its population is very young. But there are not enough jobs. The social contract is funded by oil revenue: oil also accounts for 40 percent of Saudi GDP. Saudi leadership has actually been thinking about diversification since the 1970s. The most recent push has been led by Mohammed bin Salman (MBS), the investment and tech-savvy millennial Crown Prince whom SoftBank CEO Masayoshi Son referred to as the "Bedouin Steve Jobs."[15]

One catch is that Saudi Arabia needs oil revenues to finance this transition. In Saudi Arabia, part of the financing strategy involved privatizing the most profitable company in the world: Aramco. While Aramco is an oil company, it is also a key institution in Saudi Arabia, having built much of the infrastructure in Saudi Arabia, and it plays a role in subsidizing the Saudi economy.[16] This was the largest IPO in history. Proceeds bolstered the Public Investment Fund (PIF), a sovereign wealth fund that MBS aims to scale to fund Vision 2030, the jobs and diversification plan. The PIF invests in companies like Uber and Tesla and in solar energy through an investment in the Softbank Vision Fund; it also funds small and medium-sized enterprises. The government has also rolled out a National Transformation Program—cutting public spending, reducing subsidies, and bolstering the private sector, with the goal of creating 450,000 nongovernmental jobs. All this has not been seamless—the addition of a value added tax of 5 percent on consumption of goods was tough.

Saudi Arabia needs more than oil revenue to diversify. It also needs external investment. But this vision of a sustainable future conflicts with the authoritarian style of the regime. While this may not be an operational problem—renewable energy is perfectly compatible with authoritarian governance—it's a problem for overseas investors. For example, in 2017 in the Ritz-Carlton in Riyadh, the regime held an event dubbed "Davos in the Desert." Here they granted Saudi citizenship to a robot named Sophia and launched the idea of NEOM, a futuristic sustainable city in the desert. The regime was able to conjure a vision of the future—and it was decarbonized. On the other hand, a few days later, after the foreign investors had checked out, the Ritz-Carlton became a detention center for two hundred princes, government officials, and some of Saudi's best-known businessmen. They were rounded up and imprisoned in the luxury setting and interrogated for corruption. The government "recovered" billions. This is the kind of bold power grab that makes investors nervous. And even more than the so-called "Sheikhdown," the brutal murder of dissident journalist Jamal Khashoggi casts a harsh shadow over potential investment in a greener Saudi Arabia.

While renewables could bring a more peaceful and stable international order, when viewing geopolitics through the lens of energy, this is by no means assured. There are ways that phaseout could erode stability. We could look at many places to explore this, but as an example, let's look at Russia.

You Want It to Be One Way, but It's the Other Way

"Mineral and Raw Materials Resources and the Development Strategy for the Russian Economy" doesn't sound like the most gripping Ph.D. dissertation. But this dissertation is of note because it's Vladimir Putin's. (Notably, a woman named Olga Litvinenko, daughter of the rector of the National Mineral Resources University, says that when she was a teenager staying at her father's dacha during the summer of 1997, her father wrote the entire thesis for Putin, with her help.[17]) For many years, Putin has seen energy as a way to return Russia to the status of being a great power, a *derzhava*. The concept of an energy superpower (*energeticheskaia sverkhderzhava*) was one catchword of Putinism, write geographers Stefan Bouzarovski and Mark Bassin, and this plays into narratives about regional domination.[18] Energy is seen as a strategic weapon, and in this light, China is vulnerable and Russia becomes more relevant. If you read the summary of Putin's thesis, you see a vision in which mineral wealth does it all: provides for the sustainable development of Russia's economy, allows for competing with the corporations of the West, comprises the foundation of the defensive might of the country, preserves social stability and offers a reduction in social tension, serves as a source of foreign currency, employs the population, spatially integrates Russia with neighboring states and the world community, and so on.[19]

Reality bears out the importance of the sector in Russia. The petroleum sector accounts for 40 percent of Russian government income and employs over 1.1 million, with more than 400,000 working for Gazprom.[20] It's an industry whose reach is felt around the globe. Russia has also eclipsed Saudi Arabia as China's top supplier of oil, with the new US$25 billion, 2,800-mile Eastern Siberia–Pacific Ocean oil pipeline online. China has made Rosneft a US$80 billion prepayment for oil supply delivery over the next twenty-five years.[21]

Natural gas extraction is also booming. A LNG export facility in Sabetta, on the Arctic coast, was opened in 2017, a joint venture involving Novatek, an independent Russian company, the Chinese oil company CNPC, and the French company Total. This new facility in the Yamal Peninsula was built to target Asia, because it can deliver LNG to China in just twenty days and is, according to Putin, "one more confirmation of the status of Russia as one of the world's leading energy powers."[22] But the first shipment traveled the Northern Sea Route to end up in Massachusetts during the unexpectedly cold winter of 2017, when "the arrival of Russian molecules in Boston harbor created consternation and outrage," as Daniel Yergin recounts. The Massachusetts utility had no choice but to buy the cargo that was available, and, as Yergin pointedly writes, "Massachusetts sits near the vast volumes of inexpensive gas in the Marcellus shale, which would have enabled it to avoid Russian molecules. But environmental activists and regional politicians have unwaveringly blocked construction of a new pipeline from Pennsylvania."[23] The anecdote is an uncomfortable portent of a future where successfully blocking production in one jurisdiction, without demand decline, allows production in a differently regulated area to continue.

Despite commitments by the Russian state to some level of climate action, the sector does not look poised for change. Indra Overland and Nina Poussenkova, in their 2020 book *Russian Oil Companies in an Evolving World*, identify several factors that contribute to the sector's stagnation: the denial of the significance of the shale revolution, the presence of elderly men as the top managers of the companies, and the "grab-and-run" mentality of Russian business in post-Soviet society.[24] They write:

> As a corollary to the assumption of post-Soviet short-termism, it could be argued that Russian oil companies are geared towards milking what remains of the Soviet resources and infrastructure rather than doing the demanding work needed to create something new and financially sustainable ... At times, Russia's entire post-Soviet society has seemed to be one large asset-stripping operation, selling off everything from scrap metal to human resources—and oil ... In this regard, Russia is a rentier state in a

> very literal sense: not only does it live off the easy-come, easy-go income from oil, but within the petroleum sector it is milking the infrastructure, competencies and structures inherited from the ancien régime."[25]

They ask: Are Russia's oil companies underestimating the potential of climate policy and overestimating the future demand for their products? It's hard to say—there are two conflicting visions of reality here.

Russia has a large nonhydrocarbon sector, which should make it more capable of pivoting. But much of the nonhydrocarbon sector is uncompetitive on global markets and is dependent on domestic demand, which in turn is driven by the total value of oil and gas revenues.[26] There was a program of modernization and diversification, promoted by Dmitry Medvedev, but the urgency around it dissipated during around 2010, when oil prices were at $100/barrel. Those revenues were instead put toward social welfare and rearmament. This made the collapse of oil prices a few years later even more difficult. The economic outlook grew grimmer with Western sanctions in response to Russia's annexation of Crimea and the involvement in eastern Ukraine. Plans to diversify the economy are contingent on a thriving hydrocarbons sector, argue Michael Bradshaw and colleagues. "In stark contrast to official rhetoric, Russia's current policy course is that of a country doubling down on its bet that hydrocarbons will support economic (as well as social and political) development well into the future."[27] They explain:

> There is a simple explanation for the apparent reluctance of Russian policy-makers to prepare for a world in which demand for hydrocarbons either plateaus or declines: they simply do not believe that there is a significant chance of this taking place, at least not in the next few decades.[28]

Strategic planning documents around energy, the Arctic, national security, and naval strategy make clear that many Russian policymakers see an intensification of competition over natural resources and a world characterized by resource scarcity. It is a different way of seeing. Politicians in places like the European Union, and progressives and climate professionals around the world, want to see this century as one of decarbonization. The idea of Russia as a potential Great Ecological Power (*ekologicheskaya derzhava*), with natural resources like forest carbon sinks that could be a solution to environmental problems, could complement this view. A country with the natural and human resources of Russia could be an amazing ecological leader.[29] I do not mean to foreclose this possibility, nor ignore the growing environmental activism in Russia. But it may be the other way; the way that some elements in the Russian government seem to see it. Climate politics has to prepare for both futures.

Perhaps the way to prove this vision of continued fossil energy wrong is to reduce demand for oil and gas around the world. Reduced demand should not even be considered a straightforward or complete "success," though, because if Russia's hydrocarbon reserves become stranded, people in Russia may still be stuck with an autocratic leadership and limited economic means of transformation if the government fails them. And if Russian leadership pivots from fossil fuels to say, arms, that isn't necessarily an improvement. A drop in worldwide hydrocarbon demand is not an unqualified good thing without actions that support a peaceful, healthy transition in all countries.

Here's a question: What if continued hydrocarbon production coupled with carbon capture and storage is the price to pay for peace? That is, if hydrocarbon production in Russia could be supplemented by carbon management and carbon removal, resulting in lower carbon-intensity oil and compensation for continued production, would that higher-fossil-fuel-use version of net zero be better than a lower fossil fuel version with instability? The Russian government and Russian companies are not leaning especially hard into carbon capture, utilization and storage, which allows us to avoid asking the question—for now. But Saudi Arabia is: Aramco has a whole platform around the circular carbon economy. For example, with the Abu Dhabi National Oil Company, their vision includes continued oil production through making low-carbon oil. (Their oil already naturally has a lower carbon intensity, so they have a competitive advantage, too.)

I am not suggesting that a high-fossil-fuel future is safer or more desirable than an actual end to fossil fuels. What I am suggesting is that we need a framework that addresses these tradeoffs in a genuine way. If we want to make a compelling argument for phaseout, we need to be able to speak to these geopolitical considerations.

I asked Alexandra Gillies, author of *Crude Intentions*, about this challenge of reducing demand in a synchronized way. She said, "I mean I think overall, the demand stuff matters more than the supply stuff, right? As long as there is demand for hydrocarbons and as long as the prices are robust and money can be made, I think a lot of the industry efforts to reform and walk away from producing oil in particular, aren't going to be very enthusiastic, or effective. They're going to be more about delaying tactics than they are about proactive measures, because the demand is still there and the opportunities to make money are still there."[30] Countries are responding to the continued demand, and they're making credible commercial decisions—they will do that until we're driving electric cars or taking public transport. The tricky thing about reducing demand through policies that make oil and gas more scarce is that it could drive prices up, which could make oil and gas look like a better investment.

What Gillies points to is a strategy of differentiating between producer nations and creating workable offramps given their respective situations. We need to look at where the supply is coming from. "The amount of oil that Ghana produces is important to Ghana. It's not important to the climate, really. It's just tiny. It's a drop in the bucket, compared to what the U.S., Russia and Saudi Arabia produce," and so there is an argument for Ghana being able to continue production as part of a just transition, whereas it would be different if Russia or Saudi Arabia tried to make a financing-development argument for continued production.

What about a country like Nigeria? Gillies points out that it has the building blocks for a transition, with a young population and a dynamic economy that is diverse, but has lacked the leadership necessary to bring about a reliable power supply and other measures that would unlock big parts of the nonoil economy. But one could also see an unplanned transition scenario that is devastating. Nigeria relies on Western companies to produce its oil; what if they move away from fossil fuels? "You've already got Western companies that are trying to sell off their Nigerian assets and they'll likely find buyers, so you could see kind of a race to the bottom setting in. That could be one way that decline happens—standards and revenues decline but it's not that those hydrocarbons then are left in the ground. There would be kind of an interim stage, in which another set of companies that are not as good at extracting oil will take those assets, try it and then run them into the ground, or go out of business."[31] Gillies mentions that some of this has happened already, with blocks allocated to local companies that went under when prices dropped. And this possible future is in addition to the environmental devastation in some oil-producing regions of Nigeria, which is already beyond comprehension.

I spoke with Olúfẹ́mi O. Táíwò, an expert in climate justice and reparations, about how to think about transition in Nigeria. He points out that the failure to diversify the government revenue sources in ways that would allow non-fossil development pathways is partly the failure of elites there, but also partly because of the neocolonial institutions and global order that those elites are responding to. "Asking the parts of the world that, candidly speaking, haven't contributed nearly as much to climate change as other parts of the world to abandon a path forward that they functionally have—kind of leads to the question of why they're to blame for that, would have to be justified by some kind of compensatory gain." In broader conversations around climate response, people have posed debt-for-climate swaps. "I think that's the kind of compensatory thing that would have to be on the table if you want nation states essentially in the third world to curb production of fossil fuel," Táíwò suggests. Framing that compensation in terms of, say, technology transfer for energy production, is inadequate, because we need to take into account the economic and social implications of telling a country to curb its energy production, not just the climate impacts. "Contracting the government budget by 50 percent and then installing solar panels is not the trade that a country like Nigeria would be in its right mind to make," he points out.[32]

Thinking about specific places illustrates how phasing out fossil fuels has to be a global endeavor with local nuance. We need detailed, place-specific, justice-informed roadmaps for phasing out production crafted with local experts and communities to take this discussion to the next level. It's certainly hard to think about what a person in Baltimore or Birmingham can do with this lens—we don't necessarily have the possibility or knowledge to engage in policy in Russia. But we can be more involved in the foreign policy of our own countries, as was often the case in activism in the early 2000s. For those of us living in imperialist countries—especially my fellow US readers —there's a lot of work to do here, as our government has a staggering military footprint around the globe, grounded in our tax dollars. We can also build networks with people in other countries and figure out how to stand in solidarity. And we can be more vocal in our demands for rich countries funding clean energy transition not just in our neighborhoods, but around the world.

We can also think about ways in which phaseout could be a means of promoting sustainable development and security, though there are some cautions about tying the objectives together. For example, there's been some discussion of financially compensating countries for leaving oil assets underground. The most notable case is that of Ecuador's Yasuní-ITT initiative, which was a plan to leave a billion barrels of crude oil beneath a national forest. From 2007 to 2013, President Raffael Correa advocated for the international community to provide fair compensation of $3.6 billion, which was half of the projected value of the oil. Considering that no one wants to pay for carbon, this may seem like a far-fetched request, but it had the support of many notable actors. As Benjamin Sovacool and Joseph Scarpaci relate the story, the proposal emerged from activism in the 1990s by civil society groups, called "Option One": leaving the oil permanently underground in exchange for financial compensation. The proposal was eventually championed by the Correa administration and evolved toward an offset scheme—Yasuní Guarantee Certificates. These would guarantee that the oil would stay underground, and the certificates would be commodities on the international carbon markets. What happened, though, was that the efforts raised $336 million in pledges over five years, but were only able to collect $13 million from donors; Correa eventually asserted that Ecuador has the right to develop its natural resources, rather than be "beggars sitting on a sack of gold."[33] Donors were wary of contributing because they didn't want to set the precedent that nations should be compensated for leaving oil in the ground, and they didn't want to see this become a part of future climate treaty negotiations.

Given the amount that needs to be left underground, it could be better not to establish the precedent of paying for it to stay undeveloped. (This issue comes up with regards to assigning carbon credits for avoided emissions, too.) Sovacool and Scarpaci calculate the principle applied to Saudi Arabia: if it has 90 billion barrels of proved reserves, then at $20 per ton of carbon, it could tap into a theoretical $570 billion for keeping the petroleum underground. "There was an element of perceived regressiveness to the proposal," they write, "for paying countries and communities to do things they should already be doing on their own, or opening the door to other major emitters such as Brazil (deforestation) and Saudi Arabia (crude oil and gas) asking for billions to trillions of dollars to keep their own assets stranded."[34]

What the international relations lens tells us is: it's not enough to focus on the United States or the United Kingdom or Sweden, on New York or California. Yes, actions there help, they send signals, they shape perceptions of reality and the future. But those perceptions have limits if they're not genuinely shared by state or independent oil companies and the regimes that support and rely on them. The shift in perception or international consensus doesn't necessarily change the parameters the other countries are dealing with, like the social contract in Saudi Arabia or the structure of oil companies in Russia or the domestic politics in Wyoming.

What we get from looking through the international relations lens: the ability to go beyond local just transitions toward a globally just transition.

9

Code

"Microsoft is committed to preserving Earth's ecosystems and will be creating a Planetary Computer to build an interconnected environmental network of data and tools." This line on Microsoft's website is paired with a picture of a contorted tree half-submerged in a placid lake, an afterimage of some peaceful disaster. How are we to take such statements? Is a planetary computer the logical endpoint of ubiquitous computing? The Internet of Things extended into each remote corner, sensors across the world's biomes in obscure locations transmitting data to vast datacenters through Starlink?

Visions of net zero hinge on creating a planetary computer of some sort. It's not just pure hyperbole: it's assumed, on some level, and while it may not show up packaged in a single box like a "planetary computer," you can bet that the existing platforms will have quite a central presence in it. I mean, how does anyone know when net zero happens? It's not like alarm bells go off. On a basic level, knowledge of a net-zero state requires quite a lot of computing power for monitoring emissions and removals and determining the balance.

Strategies for both deep decarbonization and carbon removal also rest on computing. Authors from institutions like Harvard, MIT, Cornell, Stanford, and Carnegie Mellon—as well as DeepMind, Google AI, and Microsoft Research—came together to write *Tackling Climate Change with Machine Learning*, a 111-page compendium of ideas.[1] Machine learning can

- Enable smart grids
- Forecast supply and demand to help determine where variable power plants should be built
- Schedule and dispatch for power generators
- Incorporate weather into renewables planning
- Accelerate materials science and discovery for solar fuels
- Create better CO_2 sorbents
- Identify and manage storage sites for captured carbon
- Utilize satellites to measure emissions
- Enable shared mobility services, freight routing, and logistics
- Synchronize vehicle-to-grid technologies that use electric vehicles for energy storage
- Enable smart buildings in smart cities
- Optimize heating and cooling of industrial facilities
- Steer consumers and purchasing firms toward climate-friendly options
- Track life cycle data of products
- Estimate the amount of carbon sequestered in a given land area
- Power precision agriculture
- Automate afforestation
- Facilitate behavioral change like nudges around energy use
- Power climate analytics for financial investment
- Predict carbon prices and help design carbon markets

What is interesting about this paper is not its comprehensiveness, but how undeveloped so many of the ideas seem—many of them are in a speculative mode, rather than describing developments that have already taken hold. Sure, there are a few charismatic efforts that people can point to, like the Climate TRACE coalition's monitoring tool that aims to tie emissions to particular polluting facilities in real-time, or the Google-supported Restor platform, which aims to offer "ecological insights at the site level," including carbon sequestration.[2] But why isn't more investment going in this direction? Why do the "planetary computer" and "AI for Earth" feel more like empty catch phrases than things that are actually being built? Probably for the same reason that a lot of cleantech has not been built: because there's not an obvious profit in monitoring earth and tracking carbon flows, yet. However, venture capital interest in "climate tech," the new "cleantech" which has been broadened to encompass this information layer, is exploding.

There are important questions to ask before this infrastructure is designed: What data, or capital, will the platforms extract along the way? Who is the subject with the carbon platform; what relations does it make or unmake? That sounds abstract, but let's go on a speculative example. Imagine that it's 2030. You want to fly somewhere. The airline offers you a recommended carbon removal certificate to compensate for your emissions—an Instagrammable shot of a mangrove forest in the Caribbean. People like you have liked this blue carbon certificate. It has glowing reviews. Someone named Jennifer from Bristol checked in at that certified location and reported that you could kayak around the mangroves, she saw three rare bird species, and there was a great vegan restaurant nearby. You click on that removal certificate. Done. You're on your way.

So is that great? How much is the platform making on your interaction? How has the certificate's appearance on the platform shaped the relationships in that community? What happens to that place in a few years, when that mangrove is sequestering all it reliably can, and it can't offer further removal certificates?

Maybe that scenario seems too far-fetched. Here's one that's closer to home. In the United States, there is a tax credit that can be claimed by capturing CO_2 and injecting it into oil wells. The tax credit is claimed by a company. They claim to have sequestered a certain amount of tons. A senator has the IRS audit them. A majority of tons were improperly claimed. Code powers the monitoring and tracking; the platforms are not open. It's a process to get those numbers; you have to be an expert to navigate that process. No one really knows what is going on with the tons of carbon dioxide.

This happened: in 2020, the Treasury Department's Inspector General for Tax Administration found that 87 percent of claims from the top ten entities for the 45Q clean air tax credit were made while the companies were not in compliance with Environmental Protection Agency (EPA) requirements. Most companies didn't submit the required monitoring, reporting, and verification plans that help EPA verify that the carbon dioxide is actually sequestered. The value of the improperly claimed credits was $893,935,025.[3] Maybe there should have been publicly accessible monitoring data, especially given that this is public money.

Take the planetary computer to its logical end under platform capitalism: every inch of the earth is mapped and monitored. Carbon flows are predicted. A red flag fire warning for a forest in Australia triggers an automatic sell-off of carbon futures; someone's bank account is crushed while they sleep. Now imagine the same platform is tracking species. Now do people. A fluctuation in the weather forecasts migrants: send more boats to Lampedusa.

All this is simultaneously hyperbolic and a logical extension of current trends. Maybe take it into what Shoshana Zuboff calls surveillance capitalism—nature's behavioral surplus fabricated into prediction products that anticipate what it will do, which are traded in behavioral futures markets.

> Automated machine processes not only know our behavior but also *shape* our behavior at scale. With this reorientation from knowledge to power, it is no longer enough to automate information flows *about us*; the goal now is to *automate us*.[4]

This births a new species of power Zuboff calls "instrumentarianism"—shaping human behavior toward others' ends.

Now instead of human beings, do birds. Now do fish. Now do trees. If all this data is blackboxed, unknowable, and used to make a profit for a mega platform, that's a horrific future—though if it was going to come to pass, you'd think it would have more hype than it does today. Perhaps the grandiosity of Silicon Valley's dreams is overestimated; one risk of studying the hyperbole of Silicon Valley is becoming hyperbolic oneself.

Let's try going with the hyperbole for a moment. Will people be compelled to accept whatever formation of this comes, as long as it is dressed in vaguely green clothing, as the price of healing the planet? Energy and climate could be a way into mass surveillance, just like health is the way into genetic engineering. As Yuval Noel Harari writes:

> Healing is the initial justification for every upgrade. Find some professors experimenting in genetic engineering or brain–computer interfaces, and ask them why they are engaged in such research. In all likelihood they would reply that they are doing it to cure disease.[5]

Why wouldn't the same thing happen with monitoring, predicting, and ultimately intervening in the environment? The rationale is to heal and restore it. Machine learning will help the planetary computer recognize species. Eventually, we end up with an artificial intelligence that understands nature and how to tend it better than we understand it; we learn about our world from the planetary computer, which is discovering new insights each day.

There is another way of reading the situation, which is that we're just in the territory of the early Internet when it comes to carbon flows and broader ecological sensing. The nascent open Internet needed to be financially sustainable as a business, and so what used to be a sphere of nonprofit and public research became an arena of for-profit and private enterprise, as Tim Hwang chronicles. It's well-documented how the state and universities laid the groundwork for the private sector Internet we know today. In his book *Subprime Attention Crisis,* which describes how the Internet's model of digital advertising is at risk of collapsing, Hwang writes about how the platforms we have were designed by entrepreneurs around the idea of advertising.[6] They adopted the New York Stock Exchange as an inspiration and template; in turn, programmatic advertising actively shaped the rules of how attention would be traded online. At the heart of the modern online system of advertising lies a mental model from finance, he explains.

Now imagine that mistake replicated for the arena where carbon flows are monitored and traded. Researchers and NGOs lay the groundwork, funded by philanthropists and governments. Hwang recites the steps of commodification: standardization, abstraction, speculation—when it comes to carbon, we're not even through that first step, but it's coming, and it's doable. If you think carbon will be tough to fully commodify, think about attention. It's even more abstract than carbon, yet we have ad impression measurement guidelines: a viewable impression, Hwang reports, involves more than 50 percent of the pixels of ad occupying the viewable space of a browser page for one continuous second after the advertising renders. Similar work around the standardization and abstraction needed for "good governance" of carbon flows can later be used to set up exchanges. If things go as they are now, Microsoft or Google will just buy up the initial startups in ecological sensing and modeling and end up owning the whole infrastructure. Why not? We could just repeat the mistakes with the Internet again.

There's a better planetary computer out there to be made. Doing so is critical for addressing climate change. Benjamin H. Bratton lays out this case powerfully in *The Terraforming*, a design manifesto for a research program underway. The starting assumptions: "the planet is artificially sentient; climate collapse mitigation and pervasive automation can converge ... 'Surveillance' of carbon flows is a good thing."[7] Indeed, "climate change" as an idea itself is an accomplishment of planetary-scale computation; he observes, it's legible and communicable as a concept because of earth systems sensing, surveillance, and modeling. Having a planetary sensing infrastructure that can recursively act upon the ecological events that it models isn't a bad thing—"We *want* our climate models that demonstrate looming systemic risk to have the kind of capacity for granular-level feedback on the ecology itself that financial models of risk have on the transactions they observe and indirectly administer," Bratton writes.

> It's not that the crisis is so big that now it's OK to acquiesce to the big panopticon, but that the form and content of the crisis makes clear that to envelope all planetary-scale sensing, indexing, and calculation technologies into a general bad category of surveillance is both intellectually lethargic and politically reactionary.[8]

Seeing planetary-scale computation through the trope of the "panopticon" would result in a politics of prevention, which isn't what we need.

Instead of rejecting planetary-scale computation and ecological sensing, we need to bring them under public control and direct them to public aims. Including phaseout.

A Tale of Two Transitions

Eventually, Big Tech and Big Oils may merge, under the domain of carbon management, into some new kind of entity. Narratives of fossil fuels and technology being sequential, as in one dynasty replacing the other, or fundamentally different, as in digital and material, keep us from seeing the possibility of a synthesis, of the energy transition becoming techified in a way that still centers fossil fuels as a key part.

We are in the midst of two transitions. One is from fossil fuels to other energy sources. The other is from legacy media and the human creative economy to digital media and artificial intelligence. As Jonathan Taplin writes:

> The last ten years have seen the wholesale destruction of the creative economy—journalists, musicians, authors, and filmmakers—wrought by three tech monopolies: Google, Facebook, and Amazon. Their dominance in artificial intelligence will extend this "creative destruction" to much of the service economy, including transportation, medicine, and retail.[9]

Both these transitions feature an entire group put out of work by technological change; scores of jobs lost along the way. The number of journalists in the United States has gone from half a million at its peak to 174,000.[10] Compare that to the mourning of coal jobs: one gets a lot more requiems and is a social problem to communally solve; the other is not. One gets a lot of technological determinism—a narrative of historical inevitability, as Adrian Daub puts it, honing in on Silicon Valley's advice to "fail better" and the smoothness of this narrative of technological evolution.[11] The other does not: the rules of creative disruption don't apparently apply to fossil fuel companies: we're supposed to empathize with them, help them continue in the face of technological disruption.

Technology is also supposed to be *beyond* resources, this next thing in a sequence: data as the new oil, the new heart of the economy; technology firms eclipse oil firms on the listings, and so on. Harari writes about how the global economy has been transformed from a material-based economy into a knowledge-based one. In the past, gold mines and wheat fields and oil wells were sources of wealth, now the main source of wealth is knowledge, and hence the profitability of war has declined.

> In 1998 it made sense for Rwanda to seize and loot the rich coltan mines of neighboring Congo, because this ore was in high demand for the manufacture of mobile phones and laptops, and Congo held 80 percent of the world's coltan reserves. Rwanda earned $240 million annually from the looted coltan ... In contrast, it would have made no sense for China to invade California and seize Silicon Valley, for even if the Chinese could somehow prevail on the battlefield, there were no silicon mines to loot in Silicon Valley.[12]

In this narrative, now we have peaceful commerce—never mind that technology still runs on child-mined coltan from the Democratic Republic of Congo.

What these narratives keep us from seeing is that these seemingly separate domains could very well merge under the domain and goal of carbon management. Artificial intelligence for Earth may help carbon monitoring become affordable and reliable. And all this data from monitored carbon needs to be stored in some cloud; platforms need to exist for exchange of emissions and removals. It's not clear whose terrain this will be, yet, but it's hard to imagine it not including Amazon, Google, and Microsoft.

Think about Microsoft's commitment to being carbon negative by 2030: it wants to remove all its emissions going back to 1975, but there are limited carbon removal opportunities on the market. So it launched a $1 billion climate innovation fund—it literally aims to create this market. Microsoft's 2021 white paper, "Carbon Removal—Lessons from an Early Corporate Purchase," states that "Microsoft and other entities committed to climate action need to take what is currently an immature market for negative emissions technologies—or carbon removal—and expand it as quickly as possible."[13] They note that a bunch of companies—Apple, Facebook, Google, Amazon—are incorporating carbon removal into their climate strategies and that Stripe and Shopify, like Microsoft, are making it a core focus. Why? Technology companies are the world's most valuable by market cap, with the exception of Aramco; they have a lot to lose from a world in climate crisis. Is the tech faction of capital finally breaking off from fossil capital and confronting climate change? Perhaps tech has reached a point of power where it can bend the direction of fossil fuels.

The Twin Challenge: Managing the Direction of Tech Platforms

Is this marriage of Big Oil and Big Tech made in hell? Or is it an opportunity? What does it have to do with the prospects of phaseout?

Phaseout can use support from tech: for the data, the capital, and the expertise. Think about code in terms of planning managed decline, that intricately synchronized dance. The plan has a layer of physical infrastructure, but it also has an information layer. And that information layer has a politics. Some of the dimensions are obviously political ("transparency"); others may present as neutral. *Why not close this plant at this time? The data shows ...* It is inevitable and good that planning be driven by data. The point is to ask questions about whose data it is, how it is created, and so on.

For an early, concrete example of how information and computing can support phaseout, consider the LA100 study done by the Los Angeles Department of Water and Power and the National Renewable Energy Laboratory (NREL), with support from academic researchers and an advisory group of community organizations.[14] The study looks at how Los Angeles could get to 100 percent renewable energy. It used NREL's supercomputer to run more than 100 million simulations; the analysis integrated electricity demand modeling, power system operations analysis, grid modeling, greenhouse gas life cycle analysis, photochemical air quality monitoring, aerial scans of rooftops to examine scenarios of consumer rooftop solar adoption, and more. So here we have public computing resources answering questions in dialogue with what public organizations want to know, and with data explorable via a public interface.

Public control and access of the information layer is the key to planning a workable, justice-centered phaseout. People ask: Why should we be able to do planning any better than nation-states did in the middle of the twentieth century? The informational context is completely different. As recently as the 1990s, people would argue that central planning was impossible because it requires an impossibly vast amount of information. Just a few decades later, we do have impossibly large amounts of information—and machine learning to make predictions to fill in gaps.

"For many progressives, the story of logistics and planning seems musty and old," write Phillips and Rozworski in *The People's Republic of Walmart*. But they ask a critical question as they survey the information landscape: "Could the plans that capitalists use every day to get goods and services into the hands of those who can pay for them be transformed to instead ensure that what we produce gets to those who need it most?" In their analysis, "Walmart is not merely a planned economy, but a planned economy on the scale of the USSR smack in the middle of the Cold War."[15] Amazon, too, is a planning device; these megacorporations indicate that the technology for planning exists.

> If today's economic system can plan at the level of a firm larger than many national economies and produce the information that makes such planning ever more efficient, then the task for the future is obvious: we must democratize and expand this realm of planning—that is, spread it to the level of entire economies, even the entire globe.[16]

There's also a role for code in monitoring. Imagine that there's not actually a decline in fossil fuels happening and that it just appears that way on the Internet. There are a few ways in which this could arise. One is that people hear so many stories about closures that it becomes a generally vague truism that production is declining—fossil fuels are socially cancelled, so they must be declining. Another driver of "false decline" could be a starry-eyed emphasis on platforms rather than content. Adrian Daub, in *What Tech Calls Thinking*, traces the de-emphasis on content back to Marshall McLuhan. Content is in a way secondary, even though platforms depend upon it—"to *create* content is to be distracted," Daub writes; "to *create* the 'platform' is to focus on the true structure of reality."[17] The platform creators are the people who become billionaires. So the tendency for developers of the Planetary Computer may be to focus on the platform, not the quality of the content. In fact, there may be a lot of shortcuts in the quality of the content. Satellite measurements and models assume carbon is stored in soils or trees, but in fact, the models may be incorrect. Yet that content—the ton assumed sequestered—is exchanged for a quantity of oil produced. Or a declared emissions reduction from a methane leak that got cleaned up isn't so verifiable on the second pass. A third potential driver of the "false decline" is straight-up fraud. Could the pressure of ambitious targets drive falsification of records? Ask Volkswagen. In theory, platforms would offer solutions for this. But sensors can be hacked; information can be entered incorrectly. So much of this comes down to the design of the system. Yet with transparent data, we can pinpoint particular polluters. Now imagine a map where you can see carbon pollution from any entity on earth. This system could extend our problems, or it could make a tremendous difference in solving them.

Imagine, too, if technology levered its social media power toward phaseout. Essentially, platforms like Facebook and Twitter have designed a behavior modification machine, or at least that's what they tell advertisers—the whole promise of the business model involves charging business for eyeballs that will eventually go and buy stuff. These platforms could in theory amplify public health information about fossil fuels. Why don't they already? Their approach to managing Covid-19 content shows they have the ability to.

Thinking about phaseout through the lens of code also helps us think about how tech helps to govern what is on the other side of ending fossil fuels. In some places, fossil fuel infrastructure may literally be replaced by data centers, given the existing grid connections; there's also the imaginary of workers entering the knowledge economy. The lens also helps us think about how workers are governed; which workers benefit from the new structures in a decarbonized world. Adrian Daub points out a gender division to the labor of platforms, where the programmers are men and the content providers are female: "men build the structures; women fill them."[18] Given how gendered both technology and fossil fuels are, this seems to be a structure that's likely to replicate without strong interventions.

It's not hard to see how the technology industry, as we know it now, could be terrible for phaseout. Consider the ways oil and gas are already entwined with big tech. The foundation of the partnership between Big Tech and Big Oil is the cloud, explains Zero Cool, a software expert who went to Kazakhstan to do work for Chevron and chronicled this in *Logic* magazine. "For Amazon, Google, and Microsoft, as well as a few smaller cloud competitors like Oracle and IBM, winning the IT spend of the Fortune 500 is where most of the money in the public cloud market will be made"—and out of the biggest ten companies in the world by revenue, six are in the business of oil production.[19] What are oil companies going to do with the cloud? Apparently, Chevron—which signed a seven-year cloud contract with Microsoft—generates a terabyte of data per day per sensor and has thousands of wells with these sensors. They can't even use all that data because of the scale of computation required. "Big Tech doesn't just supply the infrastructures that enable oil companies to crunch their data," explains Zero

Cool; they also provide analytic tools, and machine learning can help discover patterns to run their operations more efficiently. This is another reason why Big Oils need Big Tech; they have the edge when it comes to artificial intelligence/machine learning. "Why go through the effort of using clean energy to power your data centers when those same data centers are being used by companies like Chevron to produce more oil?" Zero Cool asks, also noting that one of the main reasons oil companies are interested in technology is to surveil workers.[20]

There's a paradox: technology could really help accelerate phaseout, but at the same time, technology needs to be regulated, for a variety of reasons. Will the tech industry step into this planet-saving role to avoid the growing scrutiny and backlash it's receiving? Will it be a negotiation, where we get climate stability in exchange for putting up with their concentration of power and their business models? Or can society still get help from tech on climate change, if we break up the tech industry and put it under public ownership?

"I don't think tech companies are equipped to self-regulate any more than the fossil fuel industry," researcher Safiya Umoja Noble told *Logic* magazine.[21] The techlash is changing our appraisal of the role of technology in our lives—it's just not clear, yet, what shape regulating them or doing something about them will take. Different camps are competing to tell a new story about the internet, writes *Logic* co-founder Ben Tarnoff, "one that can explain the origins of our present crisis and offer a roadmap for moving past it."[22] These analyses tend to take a liberal viewpoint on capitalism as a basically okay system that just occasionally needs a bit of state intervention, Tarnoff writes. But what we need—with regards to both fossil fuels and tech—is an understanding of what Tarnoff has called "the Luddite option," which is part of any democratic future. Facebook's power isn't just about money, Tarnoff clarifies; it also operates like an autocratic government: it has "functional sovereignty" (in Frank Pasquale's terms), given that it is the dominant media ecosystem in many countries. This is how Facebook can play a role in genocide. It's "a social network of more than two billion people ruled by a single billionaire," in Tarnoff's terms. By "Luddite option," Tarnoff doesn't mean plain refusal. "[The Luddites] saw what certain technologies were doing to them in the present tense, and took action to stop them. They weren't against technology in the abstract. They were against the relationships of domination that particular technologies enacted."

This is what is missed when thinking about a Luddite option with regards to fossil fuel technologies. Ending fossil fuels is not a technopurity argument. Nor is it just about reacting to the pollution caused and cleaning it up (in ways that are compatible with capitalism as a system that occasionally needs state intervention to run cleanly). Rather, the crux of the problem is *also* about *the relationships of domination that fossil fuel technologies allow for*. These relationships may be within developed countries, between the communities that both bear the pollution from extraction or refining and host gendered jobs in fossil fuels. Or it may be the domination that transnational companies or domestic elites have with producers in the global South. Cleaner fossil fuels is not the same thing as changing these relationships. Ending fossil fuels is not about ending pollution. It's about ending this set of devastating social relations.

Controlling the direction of Big Tech can have support on both the left and right. If we gain the capacity to direct the tech industry, we may in parallel gain the political power to direct fossil fuels, too. There may be a positive feedback loop in growing this capacity. It can run the other way: breaking up Standard Oil into thirty companies, back in 1911, is already used as a touchstone for breaking up Big Tech. (It's also a cautionary tale: those companies were all quite profitable and ended up making even more money. This may be why the December 2020 announcement of a lawsuit exploring breaking up Facebook only caused a 2 percent slip of its stock price: a broken-up Facebook might create several profitable companies.) But this breakup was buoyed by public power. In the 1870s, Zephyr Teachout relates, one in ten farmers was a Granger, a movement that demanded an economy built on supporting farmer cooperatives and breaking up big agricultural trusts. This fed a movement that was also against the railroads and Standard Oil. Today, there haven't been widespread in-the-street protests aimed at targeting tech giants—the US left, in Teachout's words, is "not seriously engaged in one of the most central arenas of the war."[23]

Do we nationalize technology platforms or break them up? Or, as Zephyr Teachout writes, why not both? "The best path forward is frequently, perhaps confoundingly, *both nationalization and decentralization*, not either/or."[24] Even after breaking up Facebook, Google, and Amazon, functions would be best designated as public utilities, she writes. Anti-monopoly is one part, though not the whole; as Gabriel Winant has pointed out, anti-monopolists are against the elite but don't necessarily address the social system that created it.[25]

One emergent demand is to run platforms as public infrastructure. Tristan Harris writes that attention utilities could be a new business classification. The platforms have created vital infrastructure, he argues, and they need a set of rules to scrutinize their recommendations, amplification, targeted ads, and content personalization.[26] So run this infrastructure publicly. "To those who see these proposals as well-neigh impossible, consider this: In the 1950s, if you said, 'We've got to get off coal,' it would have seemed impossible," Harris writes.

> We didn't have any alternative that would have produced nearly enough energy to support society. Similarly, alternatives to advertising, like subscriptions and micropayments, don't add up yet to a viable business model. But as with renewable-energy technologies, we *can* get to that point if we take the right steps now.[27]

Phillips and Rozworski also make this suggestion, with a stark warning.

> Social networks could be run as public utilities rather than as private monopolies—remember that we created public electricity or water works after the failures of nineteenth-century robber baron capitalism. One of the big questions of the twenty-first century will be, who owns and controls the data that is quickly becoming a key economic resource? Will it be the fuel for democratic planning, or instead for a new more authoritarian capitalism?[28]

Taking control of information flows can enable the planning that's needed for the energy transition. There can be a beneficial feedback loop here.

There's another reason for the "controlled demolition" of the way the Internet is working now, put forth by Tim Hwang: the structural issues at the heart of programmatic advertising present a threat. "Patiently waiting for programmatic advertising to break is an attractive position because it demands little of us, but it misses the bigger picture for a number of reasons," he writes in his book *Subprime Attention Crisis*.[29] Bubbles grow over time; the bigger the bubble, the harder the fall. "Waiting for the bubble to burst on its own deprives us of the ability to distribute the social costs of such a downturn in a just and equitable manner," and so a well-considered and structured implosion is preferable to an out-of-control collapse that could harm bystanders and created unintended damage.[30] Similarly, when it comes to fossil fuels, a planned phaseout could help us avoid a financial crisis if there is an unplanned exit from fossil fuels.

What we get from looking through the code lens: scrutiny of an emerging information infrastructure before it manifests and ideas about how to shape it.

10

Political Power

Make a list of the things that, from a public health and wellbeing point of view, need to be ended. What are your top ten? Top three?

From a climate standpoint, there are other things besides fossil fuels that need to be phased out: we can think about chemicals that harm ozone, deforestation, or industrial-scale beef.

But climate isn't the only ecological crisis. We can think about single-use plastics or pesticides: these have strong rationales for phaseout.

But let's not stop with the environment. What about carceral technologies? Facial recognition? Autonomous military robots? Nuclear arms? Even beyond the ecological crisis, there are many technologies and practices where we want to draw the line and cease developing these things.

Fossil fuels are just one part of a broader problem: How do we develop the political power to refuse and ramp down technologies and practices that are not in the public interest? How do we shape technological transitions? Guiding phaseout needs to be a capacity we develop in the Anthropocene, because it's not just fossil fuel production that is the problem. Even if it seems unlikely, we have to change in some awkward trial-by-fire capacity; there's not really any other option.

Who does the managing in a managed decline? As Sem Oxenaar and Rick Bosman point out, "If management is 'distributed' across a multitude of societal actors, as transition management implies, to what extent can governments manage a decline of fossil fuels? What would this management entail?"[1]

Writing about transition management, they suggest that the government's role is providing directional guidance and setting an end date for production, for example. There's kind of a paradox, though. They take the example of the Dutch phaseout of fossil fuels, which was a "crisis" response in the face of earthquakes. External pressure was an influential factor in pushing the regime toward the breakdown and phase-out phases of transition—and this rapid and unplanned-for reduction in government revenues from natural gas did have impacts on public finances and jobs.[2] And so they ask, can it really be a managed decline if it requires external pressure to get it going? If the crisis is the spark, is it really going to be guided and directed?

We need political power to *become* the managers, not just inform the managers of our opinions or call for managers after the worst has happened. Activists might say: obviously, that is what the whole movement is about! This lens of thinking about fossil fuels as a challenge of mobilizing democratic power will be familiar to many. I revisit it here because not everyone is used to thinking of ending fossil fuels as a project of building political power. The benefit of considering phaseout of fossil fuels as one of myriad things that need to be ended, as part of the power to guide, shape, and end our technologies, is that we have lots of examples to work from.

Plastics, Toxic Chemicals, and Tobacco: Stories of Endings

Let's consider some cases of other things that have been ended. These examples indicate that phaseout might not be all that extraordinary and is a capacity within our grasp. They are not perfectly applicable examples, but they all contain a constructive insight.

Single-use Plastics

The world's deepest plastic bag was observed on May 20, 1998, at the bottom of the Mariana Trench, 10,898 meters (36,000 feet) deep. The grainy photograph, discovered some twenty years later in a database of photos from deep-sea explorations, shows fragments of a white object and a translucent oblong shape, reminiscent of a very long used condom, on a murky background of sandy brown, which fades to deep green and black on the edges. It is a disappointing legacy—and the world realizes this.

Plastics are as ubiquitous as fossil fuels. Plastic bags are a special scourge, with estimated global consumption of 0.5–1 trillion bags per year. The interesting thing about plastic bags is that the ban is the most common form of action, above price mechanisms like levies.[3] The number of policies on plastic bags has tripled since 2010. China stated that plastic bags would be banned in all major cities, and by 2025, the restaurant industry must reduce consumption by 30 percent. Chile was the first South American country to ban plastic bags. Kenya banned plastic bags in 2017, which included distributors and producers; Morocco's law also banned import and production. France's law has a 2040 phaseout, which starts with single-use plastic plates, cups, and cotton buds and moves on to a broader range of plastics packaging and tackling fast-food toys and water bottles. A cultural norm is slowly evolving—and it's not confined to just a handful of countries.

How did this phaseout begin? Why has it met with some success so far, even while facing the organized lobbying efforts of the plastics industry?

Plastic bags are relatively new. The US oil and gas industry developed and promoted them in the 1970s, but they weren't initially adopted by consumers, and they were only offered consistently in US supermarkets in 1977, Western Europe in the 1980s, and other countries in the 1990s.[4] The backlash came relatively swiftly in developing countries, too, starting in the 1990s. Plastic bag bans started on the local level in the global South, with actors in the global North joining in anti-plastic bag norms more recently. As Jennifer Clapp and

Linda Swanston discuss, the worldwide emergence of the norm didn't involve conventional "norm entrepreneurs" at the international level, and they weren't initially networked: it was really a series of bottom-up events around the world.[5] Municipal waste collection and recycling were less established in the global South, and as Clapp and Swanston explain, new norms are more likely to take hold when the chain between cause and effect is short and clear and where the consequences cause harm—conditions that exist for plastic bags. The campaign began in Bangladesh in the early 1990s, and by 2002 strict legislation was in place banning the production, sale, and use of plastic bags in the capital Dhaka and later in the whole country. As they point out, global environmental norms are not necessarily spread from North to South or by transnational networked social movements or institutions around international agreements—they do not necessarily need an organized campaign with a global champion. NGOs in the global North did put forth organized campaigns with domestic action. The direct pressure on political actors from Greenpeace UK, for example, figured into the decision to ban many single-use plastics there.[6]

The phaseout of plastics is deeply entwined with the phaseout of fossil fuels, since plastics are also a significant contributor to warming. There's all the emissions associated with petroleum production, the energy involved with producing the plastics, and then the greenhouse gas emissions when plastics are incinerated. Global plastic production has quadrupled over the past four decades, and if this continues, greenhouse gas emissions from plastics would take up 15 percent of the global carbon budget by 2050.[7] What's worse, fossil fuel companies don't want to just continue that trend but accelerate it—demands for phasing out fossil fuels are leading some to think about pivoting to petrochemicals. Aramco, Royal Dutch Shell, BP, Total, and Exxon Mobil have all been pouring money into petrochemicals.[8]

Bans on single-use plastics provoke a whole host of questions that are salient for the phaseout of fossil fuels. First, there are real considerations around the sustainability of what's being substituted. Using new bags that need to be manufactured also has a footprint. Second, there are concerns that the focus on single-use plastics, such as bans on plastic straws, prove a distraction from dealing with larger challenges, like climate change. There is always the danger that small, relatable, local victories draw attention away from deeper structural changes in consumption patterns. Third, there are questions of how far local actions can go without an international agreement. Plastics are increasingly treated as an object of global environmental governance, but there has not been an international treaty yet. Fourth, bans can be undermined—in Rwanda, for example, there are reports of a black market for plastic bags even though there are high penalties.[9] One has to wonder what sort of black market for fossil fuels might rise up in a phaseout. Fifth, it's not clear if banning one form—like plastic bags—risks a piecemeal approach where one piece is considered "enough," attention moves on, and the whole ecosystem of plastics is not addressed. This could be an issue with fossil fuel phaseout that focuses on one fuel, like coal or oil. There are also things that make plastics an easier target for phaseout than fossil fuels: many products have easy substitutes, they're something that people can easily engage with, the damages are visual and striking. But in general, the bans on plastics around the globe illustrate how a bottom-up push with many actors can be complemented by actions by policymakers who understand the long-term threats in continued production. Whether the growing global norm will translate into ever more-ambitious action is yet to be seen.

Ozone-depleting Substances

Here's a very different example of a phaseout—an environmental problem with far fewer actors, in which international negotiations played a decisive role. In 1985, scientists using data from Antarctic research stations showed that after twenty years of steady values, ozone levels began dropping in the late 1970s. By 1984, the stratospheric ozone layer was only about 2/3 as thick as that seen in earlier measurements. This became known as the Antarctic ozone hole. The ozone layer provides universal protection from biologically harmful ultraviolet B radiation.

People had speculated since the 1970s that chlorofluorocarbons (CFCs), a class of chemicals used as fluids in refrigerators and in air conditioning, as well as propellants in aerosol spray cans, were responsible. Even though industry led a public relations campaign discrediting the link between ozone and CFCs, media attention got consumers to cut their demand for aerosol sprays in half. The 1977 Clean Air Act produced domestic regulation; the Environmental Protection Agency banned CFCs for nonessential uses in 1978. US production fell by 95 percent.[10] In this case, the public was motivated to act: they didn't want skin cancer, and aerosol sprays were pretty easy to give up.

In Europe, the story was different. It was believed that the science didn't justify the measures, the public was apathetic, and European countries went with voluntary codes rather than regulation. The United Kingdom was a CFC exporter, and many European actors wanted to preserve market dominance. These dynamics helped the United States be enthusiastic about an international agreement. Meanwhile, the science became more assertive in terms of the seriousness of the problem. In 1987, a "smoking gun" set of aircraft measurements definitively made the link between CFCs and ozone depletion. These scientific discoveries helped cement the will to act internationally and swiftly. US companies also made technological advantages in producing substitutes for CFCs and found they might have a comparative advantage in these new chemicals. When it came time to negotiate the Montreal Protocol, the

United States proposed 95 percent reductions; the result was a 50 percent cut by 1998 accompanied by a freeze on three major halons beginning in 1992. The agreement also featured loosened restrictions on developing nations and provided financial assistance to them.

Notably, the replacement chemicals that were developed for CFCs, called hydrofluorocarbons (HFCs), are a potent greenhouse gas. So countries had to agree to phase them out, too, which was challenging because the alternatives to them are much pricier. Nations like India were concerned that a rapid phaseout would make air conditioning too expensive. In 2016, the Kigali Amendment to the Montreal Protocol was adopted with the goal of HFC phaseout. It required rich countries to start cutting HFC use by 2019, China to freeze consumption by 2024, and other countries by 2028, though they can receive aid for acting more quickly. The Kigali Amendment thus was projected to avoid 0.4°C warming.[11] It has been ratified by 119 countries as of April 2021.[12] The United States under the Trump administration would not ratify it, but action on HFCs was finally enacted as part of an omnibus federal budget/coronavirus relief package at the end of 2020.

Like climate change, the depletion of the ozone layer involved global risks created by diverse nations, raised issues of intergenerational and international justice, and needed international cooperation.[13] Unlike with climate change, the phaseout of ozone-depleting substances was done through international agreement. It wasn't easy: there were ten years of policy deadlock, with several unsuccessful attempts to assess technological options.[14] The first assessments reinforced the idea that CFC cuts would be difficult and costly, but this was eventually overcome, and private-sector actors ended up participating in the assessment process to good effect, as they were anticipating regulatory action. The Montreal Protocol has been ratified by almost every country, and nations complied with their obligations. The ozone layer is expected to return to its natural level by 2050. The costs of phaseout were even lower than expected. One could regard the problem as essentially over, though atmospheric researchers measured a spike in CFC-11. This spike was traced to illegal activity in northeast China, and China is establishing new laboratories to better monitor production of ozone-depleting chemicals.[15]

It's important to understand that many experts don't see the phaseout of CFCs as a good model for the challenge of phasing out fossil fuels, because the problem structure is so different. The lesson here is not to try to replicate this template, but to note what contributed in this case. Public opinion mattered in ending the production of ozone-depleting chemicals. The leadership of large negotiators mattered too. An authoritative assessment of science mattered for moving negotiations forward, as did the dedication of informal networks of people working together on these negotiations.[16] Continued enforcement also matters.

Tobacco

With tobacco, there has been a debate as to whether there should be an "endgame"—an elimination of tobacco—or "tobacco control." In this way, it's similar to the debate around fossil fuels. Should we attempt to eliminate a corrupt industry, or would its conversion to cleaner products and more ethical business practices make it acceptable? The language is important. Public health experts have long called for an "endgame to the tobacco epidemic."[17] Tobacco control, on the other hand, implies that tobacco is here for the long haul. The goals of control should be to restrict when, where, and how it is used so that it does the least harm.[18]

Tobacco is yet another type of problem—a nonessential substance, but one entwined with advertising and culture in ways that makes it hard to end. There are a few similarities with fossil fuels, though. The companies are bad actors and are conflicted about their messages. Public health researcher Ruth

Malone tells the story of Philip Morris International writing her a personalized letter.

> One line jumped out at the end of the first paragraph, which focused largely on spin about the company's latest reinvention of itself as a responsible, transparent seller of death and disease: a reference to "acknowledging and addressing the social harms caused by our products, *including the phase out of combustible cigarettes.*" This oddly written sentence would be cause for genuine excitement if the report actually said anything specific about phasing out cigarettes, but it did not appear to do that. Nowhere could I find any plan or timetable for such a phaseout, despite the company's touting of new "heat-not-burn" products and reference to leading an effort to "replace" cigarettes with them.[19]

Current policies weren't designed for elimination of tobacco, and there was not a complete phaseout. New, less-damaging technologies (e-cigarettes or vaping) sprang up. A few years ago, people thinking about tobacco phaseout didn't imagine a network of vape shops spread across rural highways. There are also questions about the global dimensions. As policy researcher Elizabeth Smith asks, what if tobacco use phased out in the United States or United Kingdom but transnational tobacco companies continued to be based in those countries? "Will low-income and middle-income countries be left to carry on as countries with more resources solve their own tobacco problem?"[20] Even in the US, smoking is largely concentrated among people in disadvantaged communities, and many elsewhere think tobacco is no longer an issue.

Tobacco did receive an international framework convention-protocol effort, after much skepticism that such a thing could move forward. In the mid-1990s, civil society pushed for an instrument to be adopted by the United Nations. The idea for an international instrument was born at a meeting at the

UCLA Faculty Center in 1993, between public health scholar Ruth Roemer and visiting legal scholar Allyn Taylor, who had studied the constitutional authority of the WHO to promote international law. Taylor ended up writing about the framework convention-protocol approach to tobacco control as part of her dissertation.[21] There was initial resistance among WHO officials; a senior WHO official criticized the proposal as "ambitious to a fault" and emphasized that "it is important to be realistic." WHO officials recommended a nonbinding code of conduct on tobacco, but Taylor and Roemer did not think that would actually be effective, and might even be counterproductive.[22] (These dynamics will be familiar to anyone having worked in the climate space.) However, the framework found an advocate among the executive board, and a few years later a new WHO director, Gro Harlem Brundtland, had tobacco control on her priority list. The creators of the proposal also worked with the American Public Health Association and NGO agencies to get support for the convention.

The World Health Organization's Framework Convention on Tobacco Control came into force in 2005. This wasn't a quick process; it took over a decade. Smoking rates declined across 126 countries, and global cigarette consumption peaked at 5.95 trillion sticks in 2012, declining by 8 percent to 5.5 trillion in 2016.[23] The convention features a mix of obligations and recommendations. Restrictions on advertising and packaging and labeling requirements were included, as well as clean indoor air controls and legislation to combat tobacco smuggling. It also encourages price and tax measures and urges governments to require disclosure of what's in tobacco products, but it doesn't require these. Still, the treaty set a global norm.

Tobacco also faced both divestment campaigns and major lawsuits—not unlike fossil fuels. The tobacco industry in the United States sailed through 300 lawsuits over four decades, until, in the 1990s, there was a trial in Minnesota around conspiracy. The tobacco companies were defrauding consumers about the hazards of smoking, suppressing the development of safer products, and targeting children. The prosecution requested 40,000 pages of documents.[24] The companies tried to resist this by presenting 35 million pages, an effort that backfired spectacularly and ended up shifting the politics of tobacco control. Another interesting thing about the phaseout of tobacco in the United States was that it involved both demand and production: tobacco users and farmers. Tobacco had already been produced on a quota system. Legislation, funded by a settlement from tobacco companies, provided payouts to farmers.

These are just three examples of phaseouts, and they may not be the most useful or illustrative. We need an interdisciplinary assessment of many technologies and practices in which phaseout policies have been attempted. But we can ask a few important questions of these examples and think about preliminary answers.

First, when did the social narrative tip for these other examples —when did policymakers get on board and take the ideas of phaseout seriously? For each, was there a moment when it became sensible to phase them out? A moment of coming to one's senses, where a new sensibility is created? Were the advocates of phaseout even aware that they were at that moment, at the time, or did they just seem like radical thinkers? The takeaway here may be that the professional class of workers needs to take up the view of sensibility, like Ruth Roemer and Allyn Taylor were able to get their colleagues to do, or like the advocates for plastic bag bans around the world were able to do.

Second, we can look at the temporality of these phaseouts. They took at least a decade of early work before they were considered seriously, and they also have a long tail. For plastics, we don't know when the peak will be, because plastic production is not neatly monitored. For tobacco, there has been a peak and then a slower decline; for CFCs, which have a rather strict international agreement, the phaseout has had a timetable, which is largely working. People sometimes say that the Montreal Protocol worked because it only required a few big companies to agree—this may be true, but the conditions are similar to fossil fuels, which are dominated by a few companies. And—this may be critical—the companies that produced the CFCs also produced the replacement chemicals, so they still had a route to profitability. This may be raised as a "pragmatic" argument for preserving fossil fuel companies.

One key takeaway here is not to mistake the discursive peak—the moment where the narrative or norm switches over—for success. There is so much work to do on the other side, particularly if the aim is total elimination. In fact, the peak may mean lost energy for the challenge of total elimination of production. There's also a legal dimension to compliance and enforcement; black markets or illicit activity can frustrate the best of intentions.

Fossil fuels are very different than plastics, ozone-depleting chemicals, or tobacco. For one, there's already a lot of people invested in thinking about the problem a certain way, so it's not just the fossil fuel industry to be in dialogue with—it's the climate-change complex of think tanks, sustainability corporations, NGOs, and so on, who are creating policy around net zero. Second, fossil fuels are essential. Unlike plastics, tobacco, or CFCs, each person needs energy for daily things like staying warm, producing and obtaining food, and going to work. So there is potentially a deep and unspoken fear there—that phaseout could cause existential harm—which doesn't exist with other things. The fossil fuel industry has not really mobilized that fear of not having enough energy, yet. They are basically playing nice. It is still latent; most people don't think about having their electricity or gas turned off or shortages at the pump. If the fossil fuel industry does mobilize that fear, it will make phaseout far more difficult.

Political Power in Rural Areas

"I wasn't allowed to do this, but I did it anyway. I would climb up onto our roof and then climb up into the higher part. And the sunsets here are absolutely beautiful, and especially with the cornfields ... It's so pretty. Around here is flat, but once you go more towards the west I think ... it's just hill after hill of corn fields and beans and it's so pretty." I was speaking with a young woman in Ellsworth, Iowa, population 499. It's true that Iowa has plenty of cornfields. The scale of terraforming the landscape for a singular aim—corn and soy production—is staggering from the air. One might think, why not throw wind turbines in there, too? The land has already been converted for production. In the maximally renewable scenario put forth in Princeton's Net-Zero America report, that is, to reach net zero by 2050 without wind or CCS, 37 percent of Iowa must have wind or solar deployed.[25] But consider: people who live in Iowa may like it the way it is. They may find the sculpted rows of corn and soy beautiful. How do people elsewhere talk with them about this particular future? How do climate movements build political power toward ending fossil fuels in rural areas? Because winding down fossil fuels won't be possible without the support of these people.

"I believe our ability to confront metronormativity will determine our shared future," writes Xiaowei Wang in their book *Blockchain Chicken Farm and Other Stories of Tech in China's Countryside*. Metronormativity, Wang writes, is "the normative, standard idea that somehow rural culture and rural people are backward, conservative, and intolerant, and that the only way to live with freedom is to leave the country-side for highly connected urban oases." Wang writes about the urban–rural dynamic as central to globalization and about the ways we are intertwined across cities, villages, and borders. "Questioning metronormativity means demanding something outside the strict binaries of rural versus urban, natural versus man-made, digital versus physical, and remote as disengaged versus metropolitan as connected."[26] Wang's work focuses on a different context (China) and theme (technology), but the observation is still central to this: rural places enable life elsewhere and yet are viewed as peripheral.

Ending fossil fuels requires getting over metronormativity. The core–periphery dynamic needs to be changed, instead of reproduced with renewables. The political power to phase out the fossil fuel industry needs to include coalitions of people in rural areas, yet the binaries that Wang speaks of are alive on the left. I don't mean to discount the organizing done in many rural areas, which in some places has been critical. But the environmental left, at least in the United States, has largely been urban-focused when it comes to community-focused actions.

In the United States, the sense of peripherality experienced in rural areas fuels a dark populism. Racism and anti-immigrant sentiment in some rural areas can make them uninviting or hostile ground. I have heard this while talking about energy and the future with people in rural areas of the central United States. I have also found some common concerns. There's a shared skepticism of elites and a shared anger about Silicon Valley and big corporations. A rurally grounded Green New Deal, one aimed at genuine benefit in rural areas, is one starting point. An anti-monopoly, anti–Big Industry approach is another. This needs to be a core part of climate politics in the 2020s.

What we get from looking through the political power lens: a sense of the organizing work to be done and some themes to introduce.

We need an approach to phasedown that can make progress on all these fronts: the infrastructure, the cultural change, the geopolitics, the code, and the grassroots political power. There are some known policy tools, and the next section reviews these, with an invitation to consider: How well do these emergent policy tools incorporate these five dimensions of the challenge? What new ideas will be needed to complement them?

Part III. A Phaseout Toolbox for the 2020s

11

Moratoria, Bans, and Refusal to Finance

Imagine what the end of fossil fuels might actually look like: a legislative victory, a stroke of the pen, a speech, an obscure court ruling, a thousand plants going quiet. The actual end might be small and silent, but it will rest upon all kinds of work. People are rightly wondering whether this end can be brought about without violent insurrection, given the repressive tactics of states determined to preserve fossil fuel infrastructure and their seeming incapacity to change of their own accord. "At what point do we escalate? When do we conclude that the time has come to also try something different? When do we start physically attacking the things that consume our planet and destroy them with our own hands?" asks Andreas Malm in his book *How to Blow Up a Pipeline.*[1] Collective action against slavery involved violent resistance, and militancy was at the core of suffragette identity, he argues. Instead of sweeping militancy under the rug of civility, he writes, the movement should

> announce and enforce the prohibition. Damage and destroy new CO_2-emitting devices. Put them out of commission, pick them apart, demolish them, burn them, blow them up. Let the capitalists who keep on investing in the fire know that their properties will be trashed.[2]

People who might be inclined to dismiss this as rhetoric should ask themselves Malm's key question: Can business-as-usual be shaken without sabotage from a militant wing of the climate movement? The answer is not so clear.

I live in a country where gun-toting brigades like the Proud Boys have already organized around the idea that people are coming to take away their hamburgers, so I'm wary of what a backlash against physically attacking fossil fuels and the things they symbolize might look like and whom it would harm; it could create more counterviolence than change. But it also seems possible that property destruction could bring needed urgency to the climate breakdown.

In what follows, though, I concentrate on specific policy actions, which may be at first glance less interesting. To be clear, this is not to suggest that protest tactics are not important —the work by many movements around the world is part of what helped get us to this point. Can we even imagine the conversation on climate change without the Indigenous and youth-led movements engaged in direct action? There is a lot of important writing on active resistance, such as Nick Estes's work *Our History Is the Future: Standing Rock Versus the Dakota Access Pipeline, and the Long Tradition of Indigenous Resistance,* and Dina Gilio-Whitaker's *As Long as Grass Grows: The Indigenous Fight for Environmental Justice from Colonization to Standing Rock.*[3]

But also consider The Red Nation's statement on the US 2020 elections: "We don't just 'make' revolution, it has to be manufactured. It has to be organized. It will not be a continued uprising in the streets. An insurrectionist approach, while valid in its expression, is unsustainable."[4] What follows here are some ideas for what can complement this insurrectionist approach—things we can work on as the way is opening. What is politically possible is shifting, thanks to what the climate movements are already doing. This is an overview of some specific tools for ending fossil fuels. The following ideas are what tend to come up the most frequently. Executing the phaseout will take time. But we can, and must, start implementing these five ideas during the 2020s.

There's a rough sequence to these ideas, which can also be

approached as parallel tracks; none of these approaches exists in isolation from each other. Together, they form a relatively coherent approach. Begin with moratoria on exploration and then extraction; incorporate these commitments into a global agreement in which countries can receive acknowledgment for their efforts. Keep pressuring investors not to finance new projects. Phase out subsidies and route them to renewable energy instead. Buy out fossil fuel companies, and institute limits on extraction that will be ramped down each year. Turn these publicly held companies into carbon removal companies, providing jobs for workers formerly employed in fossil fuels. And provide aid to producer countries for offramps around the world. You could write this up into a tidy roadmap (and it wouldn't look too different from what some think tanks have already put forth). This is the most sensible course of action given the costs and dangers of climate change. What does it take to get it to *appear* the most reasonable?

What it will take is action on these other fronts: we need the roadmaps and the policies, but we need attention to these other facets of the challenge. The infrastructural lens illuminates the headaches that will be faced on the project level. These are manageable, and there are specific approaches that aren't covered here, such as "steel for fuel" or refinancing unpaid investments in early-retired plants with securitized or ratepayer-backed bonds. The cultural lens is more onerous: many of the ideas here, from bans to quotas to nationalization, have cultural resonances and conflict with deeply held values. There is work to be done in this space that doesn't look like "policy work" in terms of coming up with ways to understand or describe these ideas that might be consonant with people's values: it's communications work, it's relational. The geopolitical lens illustrates how solidarity and sensitivity to the dynamics in other countries is important. Thinking about phaseout as code helps us build relationships with potential allies in the technology sector to create open-source tools. And the need to build political power across rural–urban divides underscores all of this. This will be a process, but it can begin with some definitive policy actions.

Banning Exploration and Export

There are various types of bans and moratoria: bans on exploration, bans on extraction, bans on export, and bans on technologies that use fossil fuels. While a ban may sound extreme to people, bans are not as extraordinary as they might sound: all of these forms of bans have been practiced before around the world.

The most obvious, first-line form of ban is to end exploration for new oil and gas. This can be done right away, and it sends a strong signal. Nations could start with lands they manage and broaden it to lands within their jurisdiction. In the United States, the step of stopping exploration on government land would be significant, as the government administers 650 million surface acres and 2.4 billion acres of subsurface mineral rights, making it one of the world's largest energy asset managers.[5] One scientific analysis calculated that emissions from coal, oil, and gas produced on federal lands and waters accounted for 20 percent of greenhouse gas emissions in the United States in 2014.[6]

More ambitiously, a nation-state can also announce moratoria on fossil fuel exploration in their jurisdiction. France, Costa Rica, and Belize have done this. Some subnational jurisdictions have also banned production, for example on the state level (for example, New York State) or watershed level (for example, Delaware River Basin), to name a few. The Fossil Fuel Cuts Database tracking supply-side climate actions compiled by Nicolas Gaulin and Philippe Le Billion identified 106 moratoriums and bans enacted or legislated in twenty-two countries, with most in the United States, Canada, Australia, and the United Kingdom; local measures targeting natural gas extraction are one example.[7] So while production bans may seem remote to some, in a sense they are already here.

The United States actually had a different sort of ban for many years—a ban on the export of crude oil, which was imposed in the 1970s oil crisis. In the late 1970s, President Carter began phasing out energy price controls on oil, which President Reagan later abolished, but the ban on crude oil exports remained. In 2015, a congressional deal was struck to remove the ban in exchange for extending and expanding tax credits for solar and wind, with Alaskan senator Lisa Murkowski arguing that the export ban "equates to a sanctions regime against ourselves."[8] This illustrates that rather than being extraordinary, price controls and export bans have actually been commonplace.

What role would an export ban have in phaseout, if any? Greenpeace and Oil Change International calculate that reinstating the oil export ban could lead to reductions in greenhouse gas emissions of 80–181 million tons of CO_2 per year, or the equivalent of closing 19–42 coal-fired power plants. The idea is that without an export market, oil wouldn't be produced. There would also be a signaling effect: as Greenpeace analyst Tim Donaghy writes:

> Reinstating the ban would also send a strong signal to energy investors that the fossil fuel era is drawing to a close, act as a failsafe against future export-directed investments and carbon leakage, and provide a useful policy lever over emissions beyond U.S. borders.[9]

I spoke to John Noël, a senior climate campaigner at Greenpeace USA, about this. He points out that we could stop all fossil fuel combustion in the United States, and our emissions would look great, "but we still would be exporting our carbon pollution. So stopping fossil fuel exports is like a major, will be a major focus as we go forward."[10]

Banning Fossil Fuel End-user Technologies

Right now, banning internal combustion engines is all the rage. The United Kingdom, California, Canada, China, Japan, and many European countries are considering or have implemented bans. (The list may be longer by the time you read this.) Many announcements to phase out internal combustion engine (ICE) vehicles actually came in 2016 and 2017, even though progress in shifting transportation to cleaner fuels had been glacial for decades. What could explain the sudden appearance of these ambitious bans? Political scientists Jonas Meckling and Jonas Nahm, seeking to explain this wave of phaseout signaling, suggest that countries are competing for industrial leadership now that electric vehicles (EVs) have shifted from niche to mass market.[11] So these announcements arose when EVs were mature enough to scale and also compete with one another. The enthusiasm about phaseout isn't just about environmental leadership, but industrial renewal. States are setting technological agendas—as Meckling and Nahm observe, ICE phaseout announcements are the result of top-down elite politics announced in coordination with industry without bottom–up mobilization.

Gas appliances are also being banned, and this is something that can be started on local levels. Berkeley, California, became the first US city to prohibit natural gas in new buildings, in 2019. Other jurisdictions are attempting to follow suit, though notably, the first Google result you might find is from the Heritage Foundation, a right-wing think tank, arguing that it would restrict consumer choice: what about the chefs that love to cook with gas? Indeed, the California Restaurant Association sued Berkeley, arguing that "restaurants specializing in international food so prized in the Bay Area will be unable to prepare many of their specialties without natural gas."[12]

Bans on end-use applications will be thrust into the center of the culture wars. On one hand, the language and feel of a ban, especially on consumer-facing technologies, may provoke enough backlash that they actually damage the phaseout agenda. On the other hand, there may be situations where this is the only tool for the job. But the social backlash is one key reason why incentives for phaseout may be viewed as a more promising approach and one that can do the heavy lifting of the ban (for example, rebates and tax credits that make electric vehicles cheaper than ICE vehicles for a long period, rather than telling consumers that no ICE vehicles will be available). A ban can be viewed as a statement of intention that is supported by a whole bunch of other policies that are needed to make it reality. In the absence of formal government signaling and policy, investors can also announce that they won't finance upstream oil and gas—like the World Bank Group has done. These aren't technically bans but have a similar effect.

Global Coordination

What about leakage? If one country phases out or bans something, doesn't another just produce more? And can a global agreement help with navigating these issues? Phasing out fossil fuels needs a multiscalar effort.

The first reason to consider global coordination is the problem of helping producer nations transition. Global agreements can help with capacity building, technology transfer, and finance. Moratoria and bans could also become part of national pledges in the existing Paris Agreement. One 2019 assessment of the Nationally Determined Contributions (NDCs) of fifty-three fossil fuel producing nations found that only two countries had supply-side measures in their pledges.[13] Researchers Georgia Piggot and colleagues, writing in the journal *Climate Policy*, suggest several ways to mainstream phaseout into the UN Framework Convention on Climate Change: have the Intergovernmental Panel on Climate Change map phase-down pathways; include supply-side strategies into NDCs (like subsidy removal, moratoria, and production reductions); develop a framework for tracking extraction-based emissions, including fossil fuel extraction in the global stocktakes; provide financial and technical resources to developing countries for phasedown; and more.[14]

What about a fossil fuel nonproliferation treaty? This case has been made by international relations scholars Peter Newell and Andrew Simms, building on existing calls for a convention to end coal extraction.[15] The analogy is from the nuclear nonproliferation treaty, which has a three-pillar structure: nonproliferation, disarmament, and peaceful use. For fossil fuels, the first step would involve a mechanism for coordinating and verifying obligations for phaseout, with a transparent extraction-based accounting system. For Newell and Simms, the second pillar, disarmament, would include demand-reduction measures, many of which are already committed to lower than existing pledges under the Paris Agreement. The analogy of the third pillar, peaceful use, relates to how the "basic bargain" with nuclear nonproliferation: in exchange for not developing nuclear weapons, non-nuclear weapons states would be provided with assistance in technology for civil nuclear energy industries. In a climate context, they write, this would involve finance for low-carbon and non–fossil fuel clean energy and transport.

Some may wonder if the time of big, binding international treaties has passed. I talked with Justin Guay, the director of climate strategy at Australia's Sunrise Project, in November 2020. His career includes working for the Sierra Club in Washington, DC, on federal and international transition policy, as well as with foundations involved in climate philanthropy. Guay's work is largely dedicated to the permanent retirement of coal. He sees dedicated pots of funding for buying out and retiring old fossil fuel infrastructure and strong industrial policy as key: "You're going to need, absolutely need, mandatory industrial policy focused on targets and quotas and ensuring that we're building enough of the good stuff and retiring enough of the old stuff. Otherwise, we're just not going to get there."

But an international treaty? Not so much. "I think we are in a fundamentally different era," he tells me. "The whole climate regime was built decades ago off the back of the Montreal Protocol and it had its own logic and frankly, that logic just doesn't hold for the world we're in anymore." One driver of change now is the financial system. "The financial system is incredibly important for driving this transition. Where money flows is what will determine what gets built." Investors are driving a transition. "It doesn't get at the kind of nationalized oil companies or some of the state owned enterprises in places like China, Russia, and others, but they are not immune from global trends. They're not immune from market transformations. It's not as though you would see the US and Europe and other large markets radically transform, and then you would still see the equivalent of Russia making horse and buggies. It's a global marketplace. Things will shift if we're able to shift some of the biggest markets."

The Paris Agreement was important for creating a needed international norm around a certain temperature goal and energy transition—but many in the climate policy world see its role as a community event, rather than the treaty as the real end goal. "In my opinion, we have the international norm we need," Guay says. "We don't need more international negotiations and treaties. What we need is more domestic policy in the world's most important large emitters to drive the transition, because we don't need all 256 countries around the world to do this at the same time." The bulk of the emissions are in a handful of countries, and if they change their policies, it drives effects through the supply chains. "I don't think you can fault people for the architecture we created decades ago before we knew what would work or what wouldn't work," he acknowledges, "but you can fault people for continuing to kind of zombie along under the notion that what we were trying before, which still hasn't worked, will somehow magically work in the future."

Dealing with Transnational Companies

What happens if a jurisdiction places a ban on extraction when a transnational company has already invested there? This could trigger something called investor-state dispute settlement (ISDS): a system of courts that oversees international trade agreements. I asked Kyla Tienhaara, a Canadian Research Chair in Economy and the Environment at Queen's University who's written extensively on ISDS, to explain it.[16]

> Basically, there are treaties, over 2,600 treaties between countries. There are also contracts that are made between investors and states directly. These treaties and contracts allow foreign investors, so usually multi-national corporations, but also their shareholders and financial investors, to take any dispute that they have with a government over an investment to a process, an international legal process called investor-state dispute settlement. It's an arbitration process. Basically, if a government does something that negatively impacts the profitability of an investment, then that often fits under these very vague rules that are in these treaties about how government should protect investments.

The state and the investor each nominate an arbitrator, and they agree on someone to appoint the third, who might even be a lawyer acting on behalf of a corporation.

> These three people will basically review the evidence and determine whether or not the state should be liable for not protecting the investment. Generally speaking, they don't require states to actually reverse. It's very difficult, actually, to tell a state "You

> have to reverse this regulation you put in place or you have to change your approach to this investment," although in some cases, they have bordered on that. In most cases, what they just say is, "You have to pay the investor compensation for what you did."

This whole thing intersects managed decline in a critical way. The amount of compensation awarded may have no relationship to the amount the investor spent on the project; it's based on lost future profits, which are kind of impossible to determine given how oil prices fluctuate. "It also just doesn't take into account the drastic changes that need to be made," Tienhaara explains.

> We need to stop using oil altogether and arbitrators are certainly not looking at that when they're assessing the value of oil projects. They're not saying, "Well, that oil actually is not compliant with the Paris Agreement, so we shouldn't give any lost future profits after this date or anything like that." They don't consider those things. It's an insane system. Everyone, I think, the first time they hear that it exists, is first of all shocked that they've never heard about it and then taken aback at how bizarre it is, because it's completely one-sided. States can't bring cases against the investors for breaching the environmental provisions or for human rights abuses or anything like that. There's a limited scope for counterclaims, but it's not particularly effective in terms of the mechanism.

While there's a question of how often ISDS will be used, it already has been threatened—Transcanada threatened to sue through NAFTA for the cancellation of the Keystone XL pipeline—and even just the threat can trigger large compensations. This may have happened in Germany, which agreed to pay 4.35 billion euros to lignite power plant operators as a part of Germany's coal exit. A draft contract for these payouts stipulated that companies waive their right to pursue international arbitration. Tienhaara and her colleague Lorenzo Cotula looked at 257 coal power plants known to involve foreign investment and found that 75 percent of them were protected by at least one treaty with ISDS; Europe and Southeast Asia are regional hotspots.[17] Countries in the global South are particularly exposed because their infrastructure tends to be younger. "My real concern is not Germany or the Netherlands or Canada," Tienhaara tells me. "It's the countries in the global South where coal infrastructure is far more recent and therefore, a lot more value will be lost if there is a rapid exit from coal, and that also, those countries are just easier to push around because these investment arbitration cases are so expensive and the threat of having to pay out millions or hundreds of millions or billions in compensation might be enough to really put them off." The Energy Charter Treaty in particular is a multilateral treaty ratified by many states involved in the coal power industry, and it protects at least fifty-one plants exposed to the risk of asset stranding.

With ISDS, it's possible that companies could get more from the dispute process than the eventual sale of their assets, if the assets decrease in value. Disturbingly, there is now third-party funding involved in investor-state claim settlements—people who are backing the cases and betting that they will get a cut of the award, in the form of speculation. "You can actually be a bankrupt coal company and launch a dispute with these third-party funders paying the bills for you," Tienhaara tells me. How does one fight something so shadowy and dispersed? Some groups are raising awareness: the Transnational Institute, for example, has put together a guide for action around the Energy Charter Treaty.[18] Part of expanding the ability to implement bans will involve changing these obscure legal structures.

12

Ending Subsidies

Governments give fossil fuels about twice the level of support they do to renewables. Supporting fossil fuels costs governments between US$300 and $600 billion per year, according to the Organisation for Economic Co-operation and Development and the International Energy Agency. Ending these subsidies to fossil fuels is a go-to early step for phaseout. It's akin to taking one's foot off the accelerator. Why has it been so hard? It's no mystery: the simplest answer is that no government wants to be responsible for making energy cost more.

Supporting the fossil fuel industry can be done in any number of ways. Subsidies are admittedly fuzzy—as researcher Tim Rayner recommends, a first step in the problem is having a precise definition of them.[1] In plain language: Subsidies are government support to fossil fuels, from not collecting taxes on them or spending resources to support them. A standard definition includes "financial contributions by governments or other public bodies where there are direct transfers of funds, foregone or uncollected revenues, provision of goods or services, or any form of income or price support."[2] These definitions are almost intentionally vague—what was the last fossil fuel subsidy you enjoyed? What did it look or feel like? You probably weren't aware of it. It might have been a cheaper airline flight, or a break on your electricity bill, or some of your taxes going to research and development in the fossil fuel industry. I tried to figure this out in New York State, where I live, and it turns out that policymakers aren't even very aware of it—the state Senate introduced a bill in 2020 that would produce an annual, public accounting of the $1.6 billion of fossil fuel–related tax expenditures in New York State, in order to then evaluate them. It would also sunset these expenditures after five years, unless they were deemed to be in the public interest. The fact that we need a bill to produce a transparent accounting, even in one of the most climate-progressive jurisdictions, speaks to the problem. The bill is currently sitting in committee.

Generally, we can think about both producer and consumer subsidies. Producer subsidies are subsidies to upstream exploration and production of oil, gas, and coal, and they account for at least US$100 billion per year globally. Producer subsidies are often targeted toward new investments rather than ongoing production, and so they boost the investment metrics —they make investments look better than they actually are— thus locking in future production.[3] One example of this is the intangible drilling cost subsidy in the United States, which allows companies to write off costs related to drilling a well, like wages and costs of land clearing. Companies can also write down capital investments that would normally depreciate more gradually.[4] You might look at these subsidies and say: obviously, get rid of them! Yet in countries like Norway and Russia, oil producers pay high taxes, more than 70 percent on profits, which are used to fund public services.[5] If governments have to trade in decreased taxation for subsidy removal, it's not clear that this would always be a win. Even more costly are consumer subsidies, which are ways that governments try to make energy cheaper for users. Subsidies to oil products make up the largest part of this; subsidies for electricity and then natural gas make up the next largest portions.

How much of a difference would phasing out these subsidies make? Phasing out production subsidies in particular could save 37 gigatons of CO_2 from 2017 up to 2050, according to one analysis.[6] However, a subsequent analysis in *Nature* by Jessica Jewell and colleagues suggested only a modest impact on CO_2 emissions—removing all fossil fuel subsidies would reduce global energy demand by 1–4 percent by 2030, and the share of renewables would rise by less than 2 percent, making for a CO_2 emissions drop by just 1–4 percent.[7] They found that subsidy removal would have the largest impact on CO_2 emissions for Russia, the Middle East, and Latin America, and they pointed out that subsidy removal could disproportionately harm the poor in some countries. Why not more relief? One reason is that coal currently doesn't receive much subsidy, and it's the worst fuel in terms of emission.

This means that removing fossil fuel subsidies is not a magic wand—much more needs to be done. Notably, if subsidies had been redefined to include undercharging for environmental costs, "subsidies" would have totaled $5.3 trillion in 2015 (6.5 percent of global GDP) and global CO_2 emissions would have been 21 percent lower, according to a 2017 study—this goes to the issue of pricing carbon appropriately, which goes beyond subsidies.[8] Others have pointed out that integrated assessment models can't capture all the follow-on social effects from ending subsidies, like amplification effects through investment decisions.[9] In other words, part of the value of subsidy removal is being in one of those signals. When it comes to subsidies, the social logic is important, not just the carbon logic.

The reasons to end subsidies go way beyond the climate: most subsidies are socially regressive, meaning that benefits go disproportionately to middle and upper-middle income households. A 2010 IMF analysis found that 92 percent of fossil fuel consumption subsidies were realized by the top four quintiles of society.[10] And they take up revenues that can be spent on other things—some countries spend more on fossil fuel subsidies than on health or education.

Ending fossil fuel subsidies is a twofaced creature: one visage looks environmentally friendly and progressive; the other looks like encouraging austerity and powering the free market. The latter is why so many multilateral institutions support fossil fuel subsidy reform. The G20 countries pledged in 2009 to "rationalize and phase out over the medium term inefficient fossil fuel subsidies that encourage wasteful consumption"—a kind of green neoliberal rationale. Ecuador's experience is a cautionary tale: subsides for gasoline and diesel were rapidly phased out in October 2019 in an austerity package, *el paquetazo*, part of fulfilling conditions of an IMF loan. This led to drastic price spikes and protests and the reinstatement of the subsidies. Gasoline increased 25 percent; diesel doubled.[11] In July 2020, after the pandemic had lowered prices, Ecuador again tried to reform subsidies using a price-band mechanism.

But as advocates of subsidy reform point out, ways to reform subsidies that don't cause pain to people are well known. The International Institute for Sustainable Development put together "53 Ways to Reform Fossil Fuel Consumer Subsidies and Pricing," which profiles countries that have had success. The two main takeaways seem to be: make a plan for gradual action, and combine it with a safety net that offers other forms of support. For example, Egypt embraced a phased approach. In 2013, it was spending 7 percent of its GDP on fossil fuel subsidies. But supported by the World Bank, it paired this phaseout of subsidies with cash transfer programs and communications efforts. Or consider Morocco. It relies on imports for 90 percent of energy needs and saw an increase in energy demand by 60 percent from 2000 to 2011. Subsidies were helping people deal with the high costs of importing fuels. By 2012, fuel subsidies amounted to 6.6 percent of national GDP. Morocco decreased the subsidy to petroleum products from 2011 to 2016, from 5.2 billion to 1.1 billion, and increased funds for education and health insurance, keeping some subsidies for basic food and electricity. The phaseout of subsidies has been part of a planned transition to scale-up renewables across several government agencies.[12] There has been enough real-world experience to guide subsidy reform that works for communities.

13

Permission to Extract

From here, the next steps are trickier, with no clear-cut route. Production needs to be capped and ramped down. But what to call this, and how to go about it, is unclear.

The carbon budget identifies a finite number of emissions before warming targets are breached. There's been lots of theoretical writing on how to allocate these emissions, mapping out approaches to different ethical principles. The default policy right now is cap and trade, which allocates permissions to emit and is already in place in many jurisdictions worldwide. The cap generally is expressed as a yearly limit that will decline. These allowances are issued by the government, and "compliance" means giving up an allowance for each ton emitted. Generally, these are transferable. Europe, California, Quebec, and South Korea all have cap-and-trade systems. There are also discussions around establishing CO_2 quotas for the shipping industry, bringing it into the European Union's Emissions Trading System. Shipping companies would have a maximum of emissions rights and buy or sell rights from or to other shipping companies.

All of this, however, still focuses on emissions, not production. What would a policy focusing on production look like? The most familiar might be a price-based instrument. In essence, financial penalties would disincentivize production, which has the effect of leaving the allocations of remaining emissions to the market. This is familiar because governments can (and do) tax production—in the United States, more than thirty states have severance or extraction taxes, largely for fiscal rather than environmental reasons, and these taxes raise billions for state revenues.[1] This could be one way of funding a just transition program. Taxes could be tied to the carbon intensity of the fuels, so the highest carbon-intensity fuels would be phased out first. One downside of this is that it brings dependence on fossil fuels if the revenues aren't allocated with care. Another downside is that it's not particularly fair.

Another way of planning for a decrease of extraction could be allocating "permission to extract"—a set of tradable permits for extraction. This could be considered a quantity-based instrument. When it comes to real-world industrial policy, this may not be so unfamiliar. All kinds of commodities and goods have quotas, or specified limits, for production. Ironically, oil is already one of the main things under a quota system. OPEC has applied supply quotas since 1982. It has wide differences among its membership, and in fact its members often cheat on their quotas, but it is a mechanism that might be able to maintain oil prices in the face of falling demand.[2] OPEC in turn was based upon the Texas Railroad Commission, which confusingly doesn't deal much with railroads these days. Created in 1891, it is an agency that regulates energy in the state of Texas. It has set the rate of production for oil and gas fields since the 1930s—this emerged from an ethic of conservation and efficiency. Because there was an oil surplus through much of the twentieth century, it also aligned output to market demand up until the 1970s. One interesting thing about the commission is that commissioners are elected, making it a quota-setting agency that needs to answer to voters. While it is not currently in the business of managing production, this did come up in 2020 because of the coronavirus-related oil price crash. The point: in the United States and in the world, there is plenty of history and current precedent of specifying production amounts, and if it is clear that oil demand is on a path toward decline, these entities will be involved somehow.

Other substances have already been treated this way. Agricultural commodities have had quotas all around the world, though many have been phased out. Milk quotas in Europe ran from 1984 to 2015, for example, limiting dairy production. In the European Union, the last commodity to have a quota was sugar. Sugar quotas ran from 1968 to 2017. The main point was to support European producers above the world market price and to promote agricultural self-sufficiency. The way it worked was that the EU production quota of 13.5 million tons of sugar was divided between twenty member states, with price supports, while any production in excess was governed by strict rules.[3] Ending the system meant that there would be no limits on exports or imports. The sugar example is interesting because sugar is a commodity with health impacts—and thus a real question of how much is actually needed. The EU had to reckon with a debate around how the end of sugar quotas would impact public health. Would ending the quotas increase consumption of fructose syrup? But sugar demand was already decreasing within Europe for health reasons. Removing the quotas damaged the industry, which had been protected; and now some farmers are moving into other crops.

This example of the use of quotas for propping up industry is completely different from the idea of using controls on production to phase something out, but that's part of why it is interesting. A system of quotas, if too generous, could end up propping up the fossil fuel industry over time. It has also cost a lot to phase out quotas, through aid packages to farmers. The removal of both sugar and milk quotas led to overproduction and crashed prices. Interestingly, both these commodities are tied to energy. The changes in the sugar market led to changes in the biofuels market. Cheap shale gas led to further milk production, since animal feed costs drop when oil prices drop, and the cheap feed leads farmers to feed more, which means more milk production. At the same time, oil-producer countries buy less milk, making the oversupply problem worse.[4]

Are we at a point where we can build a model of these complex relationships between commodities, and everything else, to set optimum production levels given our social goals? Computationally, probably; institutionally and politically, probably not.

Quotas can be imposed for harmful substances that we know we don't want too much of. Take opioids. In the United States, the Drug Enforcement Administration (DEA) establishes production quotas for opioids. Between 1993 and 2015, it allowed aggregate production quotas for oxycodone to increase 39-fold, hydrocodone to increase 12-fold, hydromorphone to increase 23-fold, and fentanyl to increase 25-fold.[5] The DEA allowed the reckless manufacture of opioids. The SUPPORT for Patients and Communities Act (P.L. 115-271) was passed in 2018 and required transparency around quotas, as well as requiring them to be adjusted to reflect overdose deaths and public health. But legislators remained concerned that the law was not being properly followed. As they wrote in a 2020 letter:

> Approximately eleven billion opioid doses were put on the market in 2018—enough for every adult American to have a nearly three-week prescription of painkillers. As powerful painkillers are aggressively marketed and prescribed at high rates, this sheer volume of available opioids heightens the risk for illicit diversion and abuse ... We fear that the explanation provided by DEA ignores the clear connection between the staggering volumes of painkillers approved for production and the current overdose epidemic. The statute is clear that DEA must exercise its quota authority to serve as a gatekeeper and weigh the public health impact of how many opioids it allows to be sold each year in the United States.[6]

The letter notes that four out of five new heroin users began their addiction with prescription painkillers.

This is obviously a failure of quotas meant to control the oversupply of a dangerous substance, and it took thousands of deaths to bring it to attention. It brings up the question of who gets to set the quotas and how, and in what ways established industries will corrupt the process. It's not a simple story, either: the DEA did start reducing the quotas each year since 2017, since there were decreases in prescriptions, and production decreased 56 percent between 2016 and 2020. This was apparently not smooth and made for shortages in hospitals. And then a pandemic hit. In 2020, hospitals that had been facing opioid shortages didn't have supplies for treating patients on ventilators amidst Covid-19. As public health scholars wrote:

> We are concerned that should production continue to decrease as precipitously as has been the case over the past several years, all of the United States may turn into an "opioid desert." Further, we believe that the climate of opiophobia in the United States is progressively causing unnecessary harms to chronic pain sufferers, and that continued production reductions will potentially serve only to make these patients more miserable. Social media is rife with discussions of the "haves" and the "have nots," strongly suggesting that only chronic pain sufferers who are "connected" in some manner are able to receive and benefit from opioid analgesia.[7]

When it comes to fossil fuels, are quotas for production the right tools for the job, or not? The analogy between oil and addictive opioids is tired, and here I don't want to suggest that these are the same. But the example of opioid production quotas highlights some implementation challenges with quotas: you don't want a system where, given scarcity, only the well-connected get access. You also want transparency about how the quotas are set and gradual, well-communicated shifts in the quotas. But if this can be managed, allocating permissions to extract could be the key instrument in phaseout.

How should quotas be allocated? In other realms, sometimes quotas are given freely and are then tradable among users—tradable production quotas are used in fisheries, for example, where the resource is finite and it is in the interest of everyone not to deplete it. For fossil fuels, these kinds of supply restrictions can be understood as a way to avoid developing the hardest-to-get, low-energy oil (like tar sands or new deepwater projects). The restrictions are cast in terms of efficiency. As Geir B. Asheim and colleagues write, "Cost-efficient quantitative supply restrictions, such as a system of freely allocated and tradable extraction permits, render marginally profitable extraction projects unprofitable while leaving extraction of low-cost resources profitable."[8] There have also been suggestions to auction quotas, and use the funds raised for climate-related spending.

These types of discussions tend to lead to the question: What's the most equitable approach of allocating these quotas? Unfortunately, this question may be a discussion-ender. It's a paradox: we have to discuss equity, because it's primary, but it's so contentious that the quota discussion can't even get off the ground if it begins this way. Perhaps a better focus for getting started is developing institutions that can assign quotas with some legitimacy.

While this is a global matter, as we've discussed, the most obvious candidate with the power to assign permissions to extract is the nation-state. But what are the conditions in which a nation-state could legitimately assign quotas?

Some might look to a state of emergency to prompt this. Typically, calls for planning put in appeals to wartime planning, like the US War Production Board's 1942 national experiment—peacetime industries were converted to war production; materials were allocated and distributed; essential items like gas, rubber, and paper were rationed. In his essay "Planning the Planet: Geoengineering Our Way Out of and Back into a Planned Economy," Andreas Malm points out:

> any attempts to cut [greenhouse gas] emissions that would approach something like success would have to entail a significant degree of planning in the sense of governments assuming control over relevant parts of the economy—investment flows, consumption choices, trade, innovation, and so on—and steering them toward zero emissions at maximum speed. This is common knowledge, be it subconscious or conscious.[9]

He observes that when scientists call for the United Nations Framework Convention on Climate Change (UNFCCC) to be transformed into a "vanguard forum" that sets down decarbonization planning for the world, this would be something like Gosplan 2.0—the problem being that neither the UNFCCC or its member states own the means of production that would need to be repurposed for the aim of maintaining the planet's habitability. Someone would need to attend to spatial zoning and location decisions, he writes: coordinating the movement of biomass, building an extensive grid for renewables, maintaining a detailed carbon accounting system, and doing so without crashing what remains of biodiversity. Earth system scientists often call for adaptive cross-sectoral governance to avoid breaching planetary boundaries—but what does that really look like? A planned economy, as Malm points out. Yet, ironically, "the birth of the UNFCCC coincided with the death of the very idea that the economy can or should be controlled for any purpose whatsoever."[10]

In his recent book *Corona, Climate, Chronic Emergency,* Malm invokes war communism to call for planning. Scholars like Laurence Delina have written comparisons to wartime action;[11] even Bill McKibben calls for retooling the apparatus of production as in the previous world war. Peace conveys a state of bliss. "It wouldn't have worked much better if Bill de Blasio had said 'we're waging nonviolent civil disobedience against Covid-19 and ventilators are our flowers and songs'; it would barely have been intelligible," Malm comments.[12] But he doesn't use the language of war lazily: he points out the gulf between forcible deprivation and active renunciation of fossil fuels, with the potential of returning to fossil fuels when the emergency has passed. He also notes that for the Soviets, "the journey from war communism to tyranny was short to non-existent." Yet Malm also writes that,

> *during the transitional period* there is no escaping outlawing wildlife consumption and terminating mass aviation and phasing out meat and other things considered parts of the good life, and those elements of the climate movement and the left that pretend that none of this needs to happen, that there will be no sacrifices or discomforts for ordinary people, are not being honest. They are being less honest the longer the transition waits.[13]

While I broadly agree here—planning on a deep scale is needed, it will require some sacrifices or discomforts, glossing over that would be a grave mistake—the question is which language and set of metaphors will be needed for lasting economic, planetary, and cultural change. Is there even enough memory of these twentieth-century wars for these languages to be resonant? My main reservation with invoking war and crisis to enforce planning is that we need to start thinking of planning as a normal, common sense way of approaching the world, not an extraordinary one to be deployed (and then removed) during a crisis. One obvious thing that would enable an institution with the authority to plan, and set limits on production and permissions to extract, would be to put fossil fuels under public ownership.

14

Nationalize for Exit

To put stringent extraction quotas on fossil fuel companies, it may be necessary to own them. This was already an idea that was quietly and occasionally circulating after the 2008 financial crisis, when the US federal government helped out "too big to fail" institutions like banks and auto companies and also printed a whole bunch of money. "Quantitative easing for the planet," recommended Gar Alperovitz, Joe Guinan, and Thomas M. Hanna in *The Nation*, calling it "the policy weapon climate activists need."[1] Then, the coronavirus brought not just great public expenditure in times of crisis, but historically low oil prices. This coinciding of forces made it a great time to call for nationalization, and while the oil prices are back up, the mood around public spending has shifted.

But what is "nationalization," anyway? In brief, it means to put something into public ownership. There are different means of doing this—assuming the direction of an industry, seizing it, or buying a controlling stake. Much of the recent talk about nationalization is really talking about the latter—former US presidential candidate and governor of Washington Jay Inslee's climate plan talked about "buying out and decommissioning of fossil fuel assets." It's already on the level of something you can put in a serious climate plan.

A Very Brief History of Nationalization

One does not need to look very far to see nationalization—the examples are all around us. Oil and gas have lurched back and forth between public and private production. The past century has seen nearly sixty nationalizations or renationalizations.[2] Some historical context is warranted, as it still affects the politics and prospects of the idea today.

The oil industry began in the 1880s, with Rockefeller in the United States, as well as the tsarist oil industry in Baku, which involved the Nobel brothers and the French family of the Rothschilds. Russia has its own history of fossil fuel nationalization. The Bolsheviks nationalized the oil industry in 1917 during a "fuel famine." In the 1930s, Stalin's purges claimed many of the leaders and workers in the oil industry; Russia did not return to being an exporter until the late 1950s. So the development of oil production fell largely to US and European companies.

Rockefeller's tactics involved illegal means and bribery, and, in 1911, his company Standard Oil was broken into smaller companies. In the late 1940s, seven large oil companies controlled most oil reserves and provided about 90 percent of the oil traded in international production. These "Seven Sisters" consisted of Standard Oil of New Jersey (Exxon), Socony-Vacuum (Mobil), Standard Oil of California (Chevron), the Texas Company (Texaco), and Gulf—as well as the British-owned Anglo-Iranian Oil Company (British Petroleum), and the Royal Dutch/Shell group, a 60 percent Dutch and 40 percent British partnership.[3] These companies essentially established a private system of production management, which enabled the development of oil in the Middle East. Extraction was facilitated by fifty-fifty profit sharing arrangements between major oil companies and host governments. This was supported by the US government and a US tax code that granted US corporations credits for taxes paid overseas.[4]

But these exploitative relationships gradually began to see pushback. In the 1920s, oil companies made deals with Latin American governments. The relationship between oil companies and exporting countries turned into a struggle for control, and in 1938, the Mexican president announced the nationalization of the sector and the creation of the national oil company Petróleos Mexicanos. Oil Expropriation Day celebrates this. Struggles played out in many countries, especially after World War II, when there was both growing demand for oil and demands for sovereignty more generally. The creation of OPEC in 1960 further changed the balance of power in terms of countries over foreign companies. State ownership during this period was also a political issue: the rejection of imperialism. During the 1960s and 1970s, seventy-eight countries established national oil companies, and by the mid-1970s, state ownership was the norm.[5] However, resource nationalism declined during the 1980s with neoliberalism and basically stopped from 1986 to 2005,[6] which made people think that nationalization as an idea was over. Yet more recent nationalizations provoked the question of whether this is really the case. Some of the new nationalizations look more like creeping expropriation or soft nationalization—where the state takes regulatory control without claiming ownership of the property.

It Can Happen Here

People might think nationalization is something that happens in other countries "but not here." Yet even a country like the United States actually has "a long and rich tradition of nationalizing private enterprise, especially during times of economic and social crisis," as Thomas M. Hanna chronicles in "A History of Nationalization in the United States."[7] What it shows is a pattern of temporary nationalizations in times of crisis. In 1917, President Woodrow Wilson signed an executive order taking control of all railroads, which was also legislated by Congress, setting out a management plan and compensation for their former private owners. Unions mounted an effort after the war to purchase the railroads and run them using a board that involved workers, officials, and the public. This failed;

the railroads were returned to private ownership in 1920. The government also nationalized the telegraph and telephone networks, including Western Union and AT&T. Whether these should be publicly owned, as a kind of outgrowth of the postal service, was much debated at the time. They were returned to private ownership in 1919. Radio, too, was acquired by the Navy, which purchased radio installations, patents, and assets from Federal Telegraph and American Marconi—here, again, efforts to make public ownership permanent floundered, and the industry was given over to a new US-based company, Radio Company of America.

Both the Great Depression and World War II also featured nationalizations. The Tennessee Valley Authority nationalized an existing privately owned power monopoly, and in 1942, the government established an office (the Alien Property Custodian) that took control of seventeen companies worth $195 million ($3 billion in contemporary currency), which, as Hanna reports, performed very well under government ownership.

Finally, the financial crisis of 2008 and 2009 also featured strong government intervention. These recent events are sometimes held up as a precedent for action today. Sure, there were bailouts aplenty, but there were also configurations that neared nationalization. Freddie Mac and Fannie Mae were privately organized and listed on the stock market, and they were nationalized by George W. Bush—the government received a warrant to purchase 80 percent of their stock and put them into conservatorship. The government also took shares in AIG, Citigroup, and GMAC, the former financing affiliate of General Motors. And when General Motors filed for bankruptcy, the Obama administration established a new company (New GM) that was 60.8 percent owned by the federal government, with the rest held by the governments of Canada and Ontario and the United Auto Workers through their retiree health care fund. Then, Old GM's assets were transferred to New GM or sold. New GM held an IPO in 2010 in which the government's stake was reduced to 1/3, and the government sold off the rest of its shares over the next few years.

Government control of private companies is not so unusual as one might think. But if we are to take something from these efforts a century ago, is it that nationalization in a crisis may not hold? These nationalizations were temporary, a stopgap measure—despite efforts to turn things toward public ownership. The sort of nationalization we need for fossil fuels is something a bit different.

Consider this question: What can the government do once the companies are nationalized? It could determine the levels of production. It could use this to sunset operations in a managed way. Writing for the People's Policy Project, Peter Gowan argues the federal government's final intentions should be well known to investors from the beginning new shares should be devoted to a Social Energy Fund that would be chartered to use the stake in the firms to build up publicly owned renewable energy firms.[8] He also notes that while full nationalization would be preferable, the Supreme Court case law under the Takings Clause requires market value compensation for compulsory purchases—so the government would have to offer to voluntarily purchase the controlling stake. What would happen to the prices of the stocks in these companies? They would collapse, because the price of shares includes a speculative value based on the expectation of continued extraction, Gowan suggests. The signaling of this change would collapse the value, he argues, much like Obama's 2016 pledge to phase out private prisons impacted the valuation of stocks in private prison companies.

Nationalization Hinges on the Integrity of the Governments Involved

Beyond buyouts and crisis or wartime measures that use governmental authority to simply take control, there is also "back door nationalization" or "strategic nationalization," which are euphemisms to describe Putin's theft of the oil company Yukos. This requires a brief bit of background. The oil crisis in the 1970s rescued the Soviet economy, but, in the 1980s, oil revenues couldn't mask economic failures. Production fell rapidly in 1989, and the Russian republic asserted control of Soviet oil and gas assets within its territory, which put Yeltsin in charge of the oil money.[9] Oil production in the 1990s was done by private companies; gas was largely monopolized by the state-owned company Gazprom. Between 2000 and 2008, Putin brought oil under state control as well.

Mikhail Khodorkovsky chaired an independent company, Yukos, and he vastly increased production and brought its valuation up from $320 million to $21 billion and then $36 billion. Rachel Maddow tells the tale in her book *Blowout*. In 2003, Khodorkovsky began negotiating a deal with Lee Raymond and Rex Tillerson at ExxonMobil, which would give it 30 percent of Yukos. Khodorkovsky was jailed on a bunch of charges—tax evasion, fraud, and embezzlement—and sentenced to nine years. Then, Yukos got audited and stuck with a $27.5 billion dollar bill for back taxes, which wasn't payable because its liquid assets were frozen and production stalled. Next, Putin's Russian Federal Property Fund auctioned off Yukos's key subsidiary, which was bought for $9.3 billion by the "Baikalfinansgrup," a two-week-old company whose offices were above a vodka bar. State-controlled oil company Rosneft then used government funds to buy the company from the Baikalfinansgrup. "It was flat-out state-sponsored theft of a legitimate company," as Maddow puts it, and an international court at The Hague would find that the government had illegally confiscated billions from Yukos and its shareholders. Later, J.P. Morgan, Morgan Stanley, and Goldman Sachs signed up to help manage the Rosneft IPO: "To put it bluntly, Rosneft's IPO campaign ended up making the world complicit in Putin's theft of Yukos and spread the shame of it around the globe."[10] It was the sixth largest IPO in the world, at the time. Three oligarchs—Roman Abramovich, Oleg Deriapska, and Vladimir Lisin—paid 1 billion for stakes. BP, CNPC, and Petronas also bought shares worth a few billion. This is a cautionary tale of state acquisition that ended up with reprivatization—nationalization as a step toward redistribution to oligarchs. It's not clear how many governments even have the capacity to nationalize responsibility.

A government could have an actionable plan to sunset production once it acquires these companies. But how do you keep this intention to decommission the industry pure? Think about a Trump re-election scenario, or someone like him in charge of US national oil companies. That administration might end up with a mandate to produce every last drop. Alexandra Gillies, the oil corruption expert who has thought about this, likens nationalization to inviting the fox into the henhouse. "One of the problems in the U.S. is that the oil industry and politicians are too close and if you nationalize the industry, you'd actually be binding their fortunes even more closely together," she told me. There would need to be some kind of "takeover for exit" or receivership option where there are legally binding mechanisms for going out of business; otherwise the incentives would lead the government "to have an even greater interest in the well-being and profitability of these companies than they do already, which is huge."[11] Instead of a revolving door between government and industry, imagine no door.

The Prospects

You might think that buying fossil fuel companies would be ridiculously expensive, but this depends on context. Alperovitz and colleagues calculated that to buy the top twenty-five largest US-based publicly traded oil and gas companies, along with most of the remaining publicly traded coal companies, would be around $1.15 trillion (in 2017)—that still may be ballpark. Spread out over seven years, they point out, would bring it down to $200 billion a year, or less than half the US defense budget—the wars in Iraq and Afghanistan can be totaled to cost around $4–7 trillion. Or, looking at it another way, if this $1.15 trillion figure is at all accurate, buying oil and gas companies could cost the public a fraction of what the US government has spent on stimulus spending for the coronavirus (the most recent as of writing, in March 2021, was $1.9 trillion, and it follows upon rounds of $2.2 trillion, $920 billion, and more in the past year). The costs of buying out fossil fuel companies are doable relative to what we spend on other crises. The ability to pay for deep decarbonization and fossil fuel buyouts doesn't necessarily hinge on printing more money (a debate for another time). Conventional routes, like increasing taxes on the wealthy and diverting funds wasted on other budget line items, would still get the job done, at least for a country like the United States. It is not a proposal that is suited to every country. But in the US, and perhaps for a few other countries, the proposal to buy a controlling stake in fossil fuel companies deserves more serious thought. This includes robust debate on provisions that would ensure that fossil fuels actually get phased out.

Why is this not taken more seriously? Several hundred billion to get oil and gas out of the way seems cheap, compared to the damages of climate change. Perhaps it's just the resonance of the word "nationalization." Too many people have read Ayn Rand's *Atlas Shrugged*, where characters with names like Wesley Mouch, Tinky Holloway, or Bertram Scudder were going around nationalizing or "looting" everything.

If you call it public ownership and control, it's relatively popular. A 2020 Data for Progress survey asked US voters about "having the government take an ownership stake in companies that receive bailout funding" with regards to fossil fuel bailouts and coronavirus stimulus; 39 percent of people opposed this and 39 percent of people supported this; voters under the age of forty-five supported this proposal by a 21-point margin (52 percent support, 31 percent oppose).[12]

However, Sean Sweeney, the director of the International Program for Labor, Climate and Environment at City University of New York, writing in 2020 when prices were low because of the pandemic, cautions that because what we're seeing right now is a crisis of profitability of fossil fuels,

> it would therefore be a major mistake to imagine nationalization as a "clean up operation," a means of winding down the production of domestic shale gas and oil on the basis that, well, they are currently economic basket cases—so what's the problem?[13]

In fact, we are still largely dependent on fossil fuels and will be so until low-carbon energy is scaled up, so accelerating fossil fuel phaseout would mean that the United States will import more energy from overseas. If US production comes offline, global prices will rise, "and the winners will not be the climate, or workers; the winners will be the United States' current competitors." So, if nationalization will serve both workers and the climate, Sweeney writes, "we will need to accept that a phaseout of oil and gas is not a ten-year proposition."[14] Indeed, I would agree that this is a decades-long project. But it is one we need to lay the groundwork for immediately, by building up political power and institutions strong enough to carry it out.

15

Reverse Engineer

What does net zero look like when merged with phaseout? If the fossil fuel industry is allowed to write the roadmaps for net zero, we'll never know. Instead of asking "How do we end fossil fuels?" we can ask "What do we do with the fossil fuel industry?" If the answer is just "put it out of business," that's a missed opportunity. Civil society needs access to all the infrastructure, expertise, and tools. This is because we actually should remove legacy carbon from the atmosphere. It's not just a climate tech talking point. If our society has the capacity to clean up this carbon and put it back underground, what kind of people are we if we choose not to do that?

What we need to do with the fossil fuel industry is put it under public ownership and run it in reverse: transfer the skills and tools to large-scale carbon dioxide storage projects. To be clear, removing carbon from the atmosphere at a climate-significant scale means the creation of a significant new industry. First, it would involve outfitting bioenergy power plants with carbon capture equipment to run biomass carbon removal and storage. It would also involve building facilities of various scales to capture carbon directly from the air (direct air capture), as well as the renewable energy capacity to run these facilities. Creating this industry also means building new CO_2 pipelines to transport CO_2 from these facilities. Finally, it would involve drilling injection wells deep into underground rock formations and monitoring these wells. Scaling up direct air capture would also bring a surge of demand for equipment and steel—workers in trades like cement, chemicals, equipment manufacturing, and construction.[1] It is an entire value chain, a new industry from scratch—and it needs to be planned, not left to develop through markets.

Right now, the fossil fuel industry has the capital and experience to execute large-scale infrastructure projects. Geologists, petroleum engineers, chemical engineers, process technicians, and pipeline workers who are currently employed in fossil fuels could be employed in a carbon capture and removal industry. Moreover, many fossil fuel jobs are located where carbon sequestration jobs would be, because the geology for extracting carbon and replacing carbon is similar.

We should create this industry according to a progressive vision, grounded in worker demands and public ownership. Organized labor has long been calling for government support for the development of carbon capture, with unions from electricity, mining, steel, and utilities as more aligned within the Carbon Capture Coalition. In fact, climate policy will need to work with these groups—as the executive director for the Industrial Union Council at the AFL-CIO, Brad Markell, told *The Intercept*, "I don't see labor supporting any climate policy that doesn't include support for carbon capture and storage."[2] The BlueGreen Alliance, which involves building trade unions as well as service employees and the American Federation of Teachers, includes carbon capture, removal, storage, and utilization in its *Solidarity for Climate Action* plan.[3] Workers see the prospects for good, high-wage jobs in a new industry.

One way to transition the fossil fuel industry into a carbon removal industry and also join it with phaseout is a "carbon takeback requirement." The core idea is a mandatory link between production and sequestration. The proposition would require extractors and importers of fossil fuels to permanently store a percentage of the CO_2 generated by their products, called a "Carbon Take Back Obligation." The obligation would increase over time to 100 percent sequestration required; this idea is tracked on carbontakeback.org. Climate scientist Myles

Allen and colleagues argued for such a policy over a decade ago, explaining that it is a policy that could be begun today, with a small group of nations or companies that could require its fossil fuel suppliers to comply with it.[4]

But even a carbon takeback requirement doesn't get us to public ownership of carbon removal ventures, which is what we should be pushing for in tandem. Carbon removal projects could work like publicly owned utilities. Communities will need to see public benefit from these infrastructures and projects in order to get them built; public ownership is one way to do this.

Creating a publicly owned, worker-centered carbon removal industry is an urgent political project, and failing to take it up means a more dangerous climate, continued fossil fuel use, or both. But there's a real chance we won't demand better use of these technologies, because we're so understandably cynical about the companies that currently own most of the technology and expertise, and the ability to change them into something else.

I asked philosopher and climate justice expert Olúfẹ́mi O. Táíwò about the prospects of transforming the fossil fuel industry, and his answer was philosophical. "Can Shell and Exxon be made to grow social industries? Okay, I'm going to be really pedantic for a second," he explained. "In philosophy, there is this metaphor/thought experiment of the ship of Theseus. So, if you have a ship, and you took out one floorboard of the ship, would it still be the ship? If you took out two, would it still be the ship? If you took out three, if you took out four, et cetera. At some point, the question is, maybe you've removed enough parts of the ship that it's no longer recognizable as the original ship that it was."[5]

That's the difficulty in answering a question like this, Táíwò says. "If I think of the question, is the fossil fuel industry reformable, I think, well, if you nationalized it and put it in the concept of a much more democratically operated economy, and that economy was managed by the people ... and the people made decisions to manage it that are consistent with trends to defend the security of all people and society as a whole, as opposed to running on a profit-driven mentality ... but you still have the Shell logo on the truck and the equipment that is carrying out this new political will in this new political environment. Would that be in any meaningful sense the original company that we were talking about? I don't know, and I think the answer to your question depends upon that thing."

A reformed fossil fuel industry would be so different from how the current one operates, Táíwò says, that it's not clearly a continuation of the current system. "But if the question is, could there be organized human use of fossil fuel in a much different political environment, in the way that was consistent with the flourishing of all of the creatures on the planet, I don't see why the answer to that is necessarily no."

This brings up the question: When we talk about ending fossil fuels, what part of it are we ending? Are we just talking about the molecules coming out of the ground? I think most of us are talking about the exploitation, the pollution, the outrageous profits and corruption—that's what we want to end, more so than the substance itself. "Fossil fuels" becomes a shorthand for this whole system of exploitation. There's a danger that if we're so focused on the material substance, we'll miss the whole thing that needs to be changed, the social relations.

Epilogue

This is not going to be the decade in which we solve climate change. It is going to take longer than that. But it is still an absolutely critical decade. The 2020s can be the decade in which we build the capacity to get a plan together. In this decade, we can change our understanding of the problem and the work and build a new language. We can build the political power to confront this for real. We can let up on the "net-zero" talk, while specifying that we need an almost-zero version of net zero. We can focus on what to do with the fossil fuel industry, beginning with the steps laid out here: more bans on exploration and extraction, subsidy reform, expanding international coalitions and networks to develop a global approach, planning the retirement of infrastructure piece by piece. By the end of the decade, permission to extract, nationalization for exit, and reverse engineering the oil industry, as well as replacing petrochemicals with recycled hydrocarbons and biological products to fully end extraction could all be understood as real options, if we build the right coalitions.

The real poverty of net zero, as a language and problem framing, is that it strips out the broader socio-ecological crisis from focus. There's a growing understanding that the challenges we face lie inside the settler-colonial heart and mind. Technical approaches might reduce climate risks, but they won't heal the underlying rifts, the exploitation of nature and one another. All this decarbonization stuff can do is give us time to do the deep work.

Everyone can participate in phaseout, because the work here

is largely cultural, empathetic, and relational. It includes listening and creating new stories about retreat and ending and change. We already largely know how to execute a controlled demolition of fossil fuels, from a technical standpoint. There will be more roadmaps published; more scenarios, projections, and Geographic Information System analyses drawn up; more report launches and webinars. What we have to do is continue to make winding down fossil fuels mainstream, common sense—but not just among technocrats and the policy elite. We have to also help one another understand it as an opportunity.

Here are five of the main ideas from this book to discuss with your families, friends, and colleagues:

1. Net zero shifts the focus away from production. There are different versions of net zero. A language that can distinguish desirable versions from undesirable ones hasn't yet evolved. There is in fact an elegant balancing capacity to the net zero concept, and given that the capacity to decarbonize is different in different places, it could potentially be an instrument to allow some places more time to transition: removing carbon in the atmosphere in one place can mean that another place can continue to produce fertilizer and feed people. However, net zero is not enough. Without a plan for managing the decline of fossil fuels, net-zero talk and nascent carbon removal techniques are likely to be a discursive strategy for companies and countries to continue producing large amounts of fossil fuels. Net zero on its own is the wrong target of climate action. Curbing production and climate justice are both better goals.

2. Platform governance is transition governance. The danger of mistaking discourse for reality is real—with regards to net zero, but everything else in our society too. The political economy of the social media ecosystem makes us vulnerable to collective delusions. Having the everything-transition mediated by platforms that profit from polarization also means that there

are limits to the bottom-up delegitimization of fossil fuels. To address this, we need to be savvier about how we engage with these platforms. We also need to break up the technology industry and put it under public control, treating it as valuable public infrastructure. This isn't a "tech" issue; both democracy and the climate are at stake.

3. Climate advocates need to be thinking in an anticipatory mode to guide both the phaseout of fossil fuels and the construction of the new clean energy infrastructure we need through the tough parts of the 2020s. This includes preparing for a coming disenchantment with renewables as well as the arrival of low-carbon fossil fuels. A political strategy that fails to acknowledge the real-world challenges with scaling up renewables or the appeal of low-carbon fossil fuels will not work. The climate politics of the 2020s need to be in dialogue with rural areas, and they need to center health, center workers, and identify the ways in which fossil fuel phaseout and new industries can provide new choices for communities.

4. There are real tradeoffs to ending fossil fuels, but the reasons to end them are stronger, and more progress can be made toward a near-zero world when working through the tradeoffs. Fossil fuels kill people, they stifle innovation, and they contribute to corruption and oppression. If a majority of people in a place want to continue fossil fuels, is ending fossil fuels compatible with democracy? If fossil fuel production underlies social stability in a region, does ending fossil fuels risk conflict? An honest discussion acknowledges these challenges and tradeoffs head-on. The case for ending fossil fuels is strong enough to go there.

5. Phaseout needs to be approached in a multidimensional way. It requires deep cultural transformation. It requires both considering infrastructure from both place-based perspectives

and global perspectives. Phasing out fossil fuels requires a democratic planning capacity, and, in developing it, we can also develop the planning and political power to phase out other things that kill or harm people and ecosystems. We will need that capacity to make it through this century and beyond. The ability to end things is emancipatory.

Acknowledgments

Many thanks to the thinkers who made time to speak to me for this book. Thanks also to Sean Low and Peter Kareiva for comments and to Michael Thompson for encouragement. I appreciate the time of Rosie Warren and the production staff at Verso. This book, like many of my writings, would not have been possible without Laura Watson, who gifted her time for the childcare that got this manuscript finished. Gratitude to caregivers everywhere.

Notes

Introduction

1 Navajo Equitable Economy, "Navajo and Hopi Statements on the demolition of smokestacks at largest coal plant in the West," December 18, 2020, navajoequitableeconomy.org/2020/12/18/navajo-and-hopi-statements-on-the-demolition-of-smokestacks-at-largest-coal-plant-in-the-west.

2 "Time to Make Coal History," *The Economist*, December 3, 2020.

3 Ibid.

4 Rachel Morrison, "Gas Is the New Coal With Risk of $100 Billion in Stranded Assets," *Bloomberg Green*, April 17, 2021.

5 Goldman Sachs, "Carbonomics: The Future of Energy in the Age of Climate Change," redacted version, 2019.

6 Larry Fink, "A Fundamental Rethinking of Finance," Letter to CEOs, January 2020, blackrock.com/uk/individual/larry-fink-ceo-letter.

7 International Energy Agency, "World Energy Outlook 2020," 2020.

8 European Commission, Communication from the commission to the European Parliament, the Council, the European Economic and Social Committee and the Committee of the Regions: Sustainable Europe Investment Plan; European Green Deal Investment plan. COM(2020) 21 final; 2020.

9 Stockholm Environment Institute, IISD, ODI, E3G, and UN Environment Programme, *The Production Gap Report: 2020 Special Report.*

10 See rosneft.com/Investors/Rosneft_contributing_to_implementation_of_UN.

11 Danny Cullenward and David Victor, *Making Climate Policy Work*, Cambridge, UK: Polity Press, 2020.

12 Christophe McGlade and Paul Ekins, "The Geographical

Distribution of Fossil Fuels When Limiting Warming to 2°C," *Nature*, 517, 2015, 187–90.

13 BP, "Statistical Review of World Energy 2020," 69th ed., 2020.

14 International Energy Agency, "SDG7: Data and Projections," 2020.

15 "End U.S. Public Finance of Fossil Fuels," letter, March 18, 2021, foe.org/end-u-s-public-finance-of-fossil-fuels.

16 International Energy Agency, "The Oil and Gas Industry in Energy Transitions," 2020.

17 NASEO and EFI, *The 2019 U.S. Energy and Employment Report*, 2019, static1.squarespace.com.

18 International Renewable Energy Agency, "Renewable Energy Employment by Country," irena.org.

19 Jaquelin Cochran and Paul Deholm, eds., *LA 100: The Los Angeles 100% Renewable Energy Study*, Golden, CO: National Renewable Energy Laboratory, 2021.

20 Angela V. Carter and Janetta McKenzie, "Amplifying 'Keep It in the Ground' First-Movers: Toward a Comparative Framework," *Society & Natural Resources*, 33: 11, 2020, 1339–58.

21 PEAK Coalition, *The Fossil Fuel End Game: A Frontline Vision to Retire New York City's Peaker Plants by 2030*, March 2021.

22 Eric Larson et al., *Net-Zero America: Potential Pathways, Infrastructure, and Impacts,* Princeton, NJ: Princeton University, 2020.

23 Ali Alsaleh and Melanie Sattler, "Comprehensive Life Cycle Assessment of Large Wind Turbines in the US," *Clean Technologies and Environmental Policy*, 21, 2019, 887–903.

24 Kate Marsh, Neely McKee, and Maris Welsh, *Opposition to Renewable Energy Facilities in the United States,* New York: Columbia University, Sabin Center for Climate Change Law, 2021.

25 Katie Shepherd, "Rick Perry says Texans would accept even longer power outages 'to keep the federal government out of their business,'" *The Washington Post*, February 18, 2021.

26 H.R. 133, Consolidated Appropriations Act, 2021, congress.gov/bill/116th-congress/house-bill/133.

27 Aramco, "The circular carbon economy," aramco.com/en/making-a-difference/planet/the-circular-carbon-economy.

28 Kevin Adler, "Oxy Unit Delivers First Shipment of CO2-neutral Oil to India," IHS Markit, February 2, 2021.

29 Peter Fox-Penner, *Power After Carbon: Building a Clean, Resilient Grid*, Cambridge, MA: Harvard University Press, 2020.
30 International Energy Agency, *The Future of Petrochemicals*. IEA, 2018.

1 How to Not Say the F Word

1 H.Res.109, "Recognizing the Duty of the Federal Government to Create a Green New Deal," 2019, congress.gov/bill/116th-congress/house-resolution/109.
2 United Nations, Paris Agreement, 2015, unfccc.int/sites/default/files/english_paris_agreement.pdf. European Commission, "A European Green Deal," ec.europa.eu/info/strategy/priorities-2019-2024/european-green-deal_en.
3 Indigenous Environmental Network, "Talking Points on the AOC-Markey Green New Deal (GND) Resolution," February 7, 2019.
4 Goldman Sachs, "Carbonomics: The Future of Energy in the Age of Climate Change," December 11, 2019.

2 Inventing Net Zero

1 Megan Darby, "Net zero: the story of the target that will shape our future", Climate Home News, September 16, 2019.
2 United Nations Climate Change, Paris Agreement, 2015, unfccc.int/sites/default/files/english_paris_agreement.pdf.
3 Eva Lövbrand and Johannes Stripple, "Making Climate Change Governable: Accounting for Carbon as Sinks, Credits and Personal Budgets," *Critical Policy Studies*, 5: 2, 2001, 187–200.
4 Eva Lövbrand and Johannes Stripple, "The Climate as Political Space: On the Territorialisation of the Global Carbon Cycle," *Review of International Studies*, 32: 2, 2006, 217–35.
5 United Nations Framework on Climate Change, Kyoto Protocol to the United Nations Framework Convention on Climate Change, 1997, unfccc.int/sites/default/files/resource/docs/cop3/l07a01.pdf.
6 "Prices in the World's Biggest Carbon Market Are Soaring," *The Economist*, February 27, 2021.
7 David Hone, *Putting the Genie Back: Solving the Climate and Energy Dilemma*, Bingley, UK: Emerald Publishing, 2017.
8 Stephen Pacala and Robert Socolow, "Stabilization Wedges: Solving the Climate Problem for the Next 50 Years with Current Technologies," *Science*, 305: 5686, 2004, 968–72.

9 Steven J. Davis, Long Cao, Ken Caldeira, and Martin I. Hoffert, "Rethinking Wedges," *Environmental Research Letters*, 8, 2013.

3 What's Truly "Hard to Decarbonize"?

1 David Hone, *Putting the Genie Back: Solving the Climate and Energy Dilemma*, Bingley, UK: Emerald Publishing, 2017.

2 Ibid.

3 Abdullah F. Alarfaj et al., "Decarbonizing US Passenger Vehicle Transport under Electrification and Automation Uncertainty Has a Travel Budget," *Environmental Research Letters*, 15, 2020.

4 Dustin Mulvaney, *Solar Power: Innovation, Sustainability, and Environmental Justice,* Oakland: University of California Press, 2019.

5 Ibid.

6 Gregory Nemet, *How Solar Energy Became Cheap*, New York: Routledge, 2019.

7 International Energy Agency, *World Energy Outlook 2020,* iea.org.

8 Colin Cunliff, "An Innovation Agenda for Deep Decarbonization: Bridging Gaps in the Federal Energy RD&D Portfolio," Information Technology & Innovation Foundation, 2018.

9 Taner Şahin, "Capacity Renumeration Mechanism: Regulatory Tools for Sustaining Thermal Power Plants and the EU Energy Transition," in *The Palgrave Handbook of Managing Fossil Fuels and Energy Transitions*, eds. Geoffrey Wood and Keith Baker, Cham, Switzerland: Springer Nature, 2020.

10 Paul Dorfman, "The Long Goodbye to the Nuclear Monument," in *The Palgrave Handbook of Managing Fossil Fuels and Energy Transitions.*

11 Ibid.

12 World Bank, *The Growing Role of Minerals and Metals for a Low Carbon Future*, 2017, elibrary.worldbank.org/doi/abs/10.1596/28312.

13 Jordy Lee, Morgan Bazilian, B. Sovacool, and S. Greene, "Responsible or Reckless? A Critical Review of the Environmental and Climate Assessments of Mineral Supply Chains," *Environmental Research Letters*, 15, 2020.

14 Cunliff, "An Innovation Agenda for Deep Decarbonization."

15 S. Julio Friedmann, Zhiyuan Fan, and Ke Tang, *Low-Carbon Heat Solutions for Heavy Industry: Sources, Options, and Costs*

Today, Columbia University, SIPA Center on Global Energy Policy, 2019, energypolicy.columbia.edu/sites/default/files/file-uploads/LowCarbonHeat-CGEP_Report_100219-2_0.pdf.

16 Shell, *A Better Life with a Healthy Planet: Pathways to Net-Zero Emissions,* shell.com.

17 Giorgios Kallis, *In Defense of Degrowth: Opinions and Minifestos* by Giorgos Kallis, edited by Aaron Vansintjan, Uneven Earth Press, 2018; Giorgios Kallis, Susan Paulson, Giacomo D'Alisa and Federico Demaria, *The Case for Degrowth*, Cambridge: Polity, 2020; Leigh Phillips, *Austerity Ecology and the Collapse Porn Addicts: In Defense of Growth, Progress, Industry and Stuff,* Winchester, UK: Zero Books, 2015; Jason Hickel, *Less is More: How Degrowth will Save the World*, New York: Random House, 2020; Kate Soper, *Post-Growth Living: For an Alternative Hedonism*, London: Verso, 2020.

4 Creating Negative Emissions

1 Oil and Gas Climate Initiative, *Scaling up Action: Aiming for Net Zero Emissions*, September 2019, euagenda.eu/upload/publications/ogci-annual-report-2019.pdf.

2 Bronson W. Griscom et al., "Natural Climate Solutions," *PNAS*, 114: 44, 2017, 11645–50.

3 Ibid.

4 Giulia Realmonte et al., "An Inter-model Assessment of the Role of Direct Air Capture in Deep Mitigation Pathways," *Nature Communications*, 10: 3277, 2019.

5 Nathalie Berny and Christopher Rootes, "Environmental NGOs at a Crossroads?" *Environmental Politics*, 27: 6, 2018.

6 Laura Marsh, "The Flaws of the Overton Window Theory," *The New Republic*, 2016.

7 Ben Burgis, "Let's Stop Talking About the 'Overton Window,'" *Jacobin*, 2020.

5 Why We Need a Planned Ending for Fossil Fuels

1 Christophe McGlade and Paul Ekins, "The Geographical Distribution of Fossil Fuels When Limiting Warming to 2°C," *Nature*, 517, 2015, 187–90.

2 Karn Vohra et al., "Global Mortality from Outdoor Fine Particle Pollution Generated by Fossil Fuel Combustion: Results from GEOS-Chem," *Environmental Research*, 195, 2021.

3 Noel Healy, Jennie C. Stephens, and Stephanie A. Malin, "Embodied Energy Injustices: Unveiling and Politicizing the Transboundary Harms of Fossil Fuel Extractivism and Fossil Fuel Supply Chains," *Energy Research & Social Science*, 48, 2019, 219–34.

4 A. Farrow, K. A. Miller, and L. Myllyvirta, *Toxic Air: The Price of Fossil Fuels*, Seoul: Greenpeace Southeast Asia, 2020.

5 Mark Olade and Ryan Menezes, "The Toxic Legacy of Old Oil Wells: California's Multibillion-dollar Problem," *LA Times*, February 6, 2020.

6 Refinitiv, "Edited Transcript: Q3 2020 Occidental Petroleum Corp Earnings Call," November 10, 2020.

7 Refinitiv, "Edited Transcript."

8 House Committee on Energy & Commerce, "Hearing on 'Clearing the Air: Legislation to Promote Carbon Capture, Utilization and Storage,'" February 6, 2020, energycommerce.house.gov

9 Jimmy Carter, "A Crisis of Confidence (1976)," in *American History Through Its Greatest Speeches: A Documentary History of the United States*, ed. Courtney Smith, Santa Barbara, CA: ABC-CLIO, 2017.

10 Meg Jacobs, *Panic at the Pump: The Energy Crisis and the Transformation of American Politics in the 1970s*, New York: Hill and Wang, 2016.

11 Robert Stobaugh and Daniel Yergin, *Energy Future: A Report of the Energy Project at the Harvard Business School*, New York: Random House, 1979.

12 Barry Commoner, *The Poverty of Power: Energy and the Economic Crisis*, New York: Alfred A. Knopf, 1976.

13 Atomic Energy Commission, "The Nation's Energy Future: A Report to Richard M. Nixon President of the United States," 1973.

14 Alfred Eggers, *Subpanel IX: Solar and Other Energy Sources*, 1973, osti.gov.

15 Richard Nixon, "Transcript of President's Address on the Energy Situation," November 8, 1973, nytimes.com.

16 Geir B. Asheim et al., "The Case for a Supply-side Climate Treaty," *Science*, 365: 6451, 2019.

17 Dina Rosenberg and Georgy Tarasenko, "Innovation for Despots? How Dictators and Democratic Leaders Differ in Stifling Innovation and Misusing Natural Resources across 114

Countries," *Energy Research & Social Science*, 68, 2020.

18 Michael L. Ross, "What Have We Learned about the Resource Curse?" *Annual Review of Political Science*, 18, 2015, 239–59.

19 Alexandra Gillies, *Crude Intentions: How Oil Corruption Contaminates the World*, New York: Oxford University Press, 2020.

20 Thea Riofrancos, *Resource Radicals: From Petro-Nationalism to Post-Extractivism in Ecuador*. Durham, NC: Duke University Press, 2020.

21 Rachel Maddow, *Blowout: Corrupted Democracy, Rogue State Russia, and the Richest, Most Destructive Industry on Earth*, New York: Crown, 2019.

22 Robert Vitalis, *Oilcraft: The Myths of Scarcity and Security That Haunt U.S. Energy Policy*, Stanford, CA: Stanford University Press, 2020.

23 Daniel Yergin, *The New Map: Energy, Climate and the Clash of Nations*, New York: Penguin, 2020.

24 Asheim et al., "The Case for a Supply-side Climate Treaty."

25 Jacobs, *Panic at the Pump*.

26 Stefan Kipfer, "What Colour Is Your Vest? Reflections on the Yellow Vest Movement in France," *Studies in Political Economy: A Socialist Review*, 100: 3, 2019, 209–31.

27 Andreas Goldthau and Kirsten Westphal, "Why the Global Energy Transition Does Not Mean the End of the Petrostate," *Global Policy*, 10: 2, 2019.

6 Culture

1 Stephanie LeMenager, *Living Oil: Petroleum Culture in the American Century*, Oxford: Oxford University Press, 2014, 102.

2 Thomas Princen, Jack P. Manno, and Pamela L. Martin, eds., *Ending the Fossil Fuel Era*, Cambridge, MA: MIT Press, 2015, 12.

3 Matthew Huber, *Lifeblood: Oil, Freedom, and the Forces of Capital*, Minneapolis: University of Minnesota Press, 2013, 19, 26, 169.

4 Cara New Daggett, "Petro-masculinity: Fossil Fuels and Authoritarian Desire," *Millennium: Journal of International Studies*, 47(1), 2018, 28 and Cara New Daggett, *The Birth of Energy: Fossil Fuels, Thermodynamics, and the Politics of Work*, Durham, NC: Duke University Press, 2019.

5 Princen, *Ending the Fossil Fuel Era*.

6 Lee Vinsel and Andrew R. Russell, *The Innovation Delusion*, New York: Currency, 2020, 11.
7 Ibid., 11.
8 Ibid., 14.
9 Richard Toye, *The Labour Party and the Planned Economy, 1931–1951*, Suffolk, UK: Boydell & Brewer, 2003.
10 Leigh Phillips and Michal Rozworski, *The People's Republic of Walmart: How the World's Biggest Corporations Are Laying the Foundation for Socialism*, New York: Verso, 2019; Andreas Malm, *Corona, Climate, Chronic Emergency*, New York: Verso Books, 2020.
11 Phillips and Rozworski, *The People's Republic of Walmart*, 13.
12 Toye, *The Labour Party and the Planned Economy*, 3.
13 Ibid.
14 Bishwapriya Sanyal, "Hybrid Planning Cultures: The Search for the Global Cultural Commons," in *Comparative Planning Cultures*, ed. Bishwapriya Sanyal, New York: Taylor & Francis, 2005, 6.
15 Phillips and Rozworski, *The People's Republic of Walmart*, 213.
16 Ibid.
17 Kian Goh, "Planning the Green New Deal: Climate Justice and the Politics of Sites and Scales," *Journal of the American Planning Association*, 86: 2, 2020, 193.

7 Infrastructure

1 Paul Schnell, interview via phone, December 21, 2020.
2 James Wiliams et al, *Pathways to Deep Decarbonization in the United States*, Deep Decarbonization Pathways Project, 2014.
3 Energy Information Administration, "Electricity Explained," eia.gov/energyexplained/electricity/electricity-in-the-us-generation-capacity-and-sales.php, updated March 18, 2021.
4 International Energy Agency, *World Energy Outlook 2020*, iea.org.
5 Dawud Ansari and Franziska Holz, "Between Stranded Assets and Green Transformation: Fossil-fuel-producing Developing Countries towards 2055," *World Development*, 130, 2020.
6 Kyra Bos and Joyeeta Gupta, "Stranded Assets and Stranded Resources: Implications for Climate Change Mitigation and Global Sustainable Development," *Energy Research & Social Science*, 56, 2019.
7 Ibid.

8 International Renewable Energy Agency, *Perspectives for the Energy Transition: Investment Needs for a Low-carbon Energy System*, 2017, irena.org.
9 Kyla Tienhaara and Lorenzo Cotula, *Raising the Cost of Climate Action? Investor-State Dispute Settlement and Compensation for Stranded Fossil Fuel Assets*, London: International Institute for Environment and Development, 2020.
10 Emily Grubert, "Fossil Electricity Retirement Deadlines for a Just Transition," *Science*, 370: 6521, 2020.
11 Ibid.
12 Peter Fox-Penner, *Power After Carbon: Building a Clean, Resilient Grid*, Cambridge, MA: Harvard University Press, 2020.
13 Paul Bodnar et al., *How to Retire Early: Making Accelerated Coal Phaseout Feasible and Just,* Basalt, CO: Rocky Mountain Institute, 2020.
14 Bos and Gupta, "Stranded Assets and Stranded Resources."

8 Geopolitics

1 Simonetta Spavieri, "A First Estimation of Fossil-Fuel Stranded Assets in Venezuela Due to Climate Change Mitigation," *IAEE Energy Forum*, Fourth Quarter, 2019.
2 Ibid.
3 Grzegorz Peszko, Amelia Midgley, Dimitri Zenghelis and John Ward, "Diversification and Cooperation in a Decarbonizing World: Climate Strategies for Fossil-fuel Dependent Countries," *Development and a Changing Climate*, 2020, blogs.worldbank.org.
4 Andreas Goldthau and Kirsten Westphal, "Why the Global Energy Transition Does Not Mean the End of the Petrostate," *Global Policy*, 10: 2, 2019.
5 Peszko et al., "Diversification and Cooperation in a Decarbonizing World 2020."
6 Thijs Van de Graaf, "Battling for a Shrinking Market: Oil Producers, the Renewables Revolution, and the Risk of Stranded Assets," in *The Geopolitics of Renewables*, ed. Daniel Scholten, Delft, The Netherlands: Springer, 2018, 97–121.
7 Dieter Helm, *Burnout: The Endgame for Fossil Fuels*, New Haven, CT: Yale University Press, 2018.
8 Daniel Yergin, *The New Map: Energy, Climate and the Clash of Nations,* New York: Penguin, 2020, 111.

9 Yergin, *The New Map*.
10 Dawud Ansari and Franziska Holz, "Between Stranded Assets and Green Transformation: Fossil-fuel-producing Developing Countries towards 2055," *World Development*, 130, 2020.
11 National Editorial, "Sheikh Mohammed bin Zayed's inspirational vision for a Post-oil UAA," *National News*, February 10, 2015.
12 ADNOC, "Low Carbon Oil to Play Central Role in the Energy Transition," press release, March 3, 2021, adnoc.ae/news-and-media/press-releases/2021/low-carbon-oil-to-play-central-role-in-the-energy-transition.
13 Thijs Van de Graaf and Aviel Verbruggen, "The Oil Endgame: Strategies of Oil Exporters in a Carbon-constrained World," *Environmental Science & Policy*, 54, 2015, 456–62.
14 Michael Bradshaw, Thijs Van de Graaf, and Richard Connolly, "Preparing for the New Oil Order? Saudi Arabia and Russia," *Energy Strategy Reviews*, 26, 2019.
15 Bradley Hope and Justin Scheck, *Blood and Oil: Mohammed Bin Salman's Ruthless Quest for Global Power,* New York: Hachette, 2020.
16 Hope and Scheck, *Blood and Oil.*
17 Dmitry Volchek and Robert Coalson, "Cut-And-Paste Job: 'My Father Wrote Putin's Dissertation,'" Radio Free Europe / Radio Liberty, March 7, 2018, rferl.org.
18 Stefan Bouzarovski and Mark Bassin, "Energy and Identity: Imagining Russia as a Hydrocarbon Superpower," *Annals of the American Association of Geographers*, 101: 40, 2011, 783–94.
19 V.V. Putin, trans. Thomas Fennell, "Putin's Thesis (Raw Text)," *The Atlantic*, August 20, 2008.
20 Indra Overland and Nina Poussenkova, *Russian Oil Companies in an Evolving World*, Northampton, MA: Edwin Elgar, 2020.
21 Yergin, *The New Map,* 118.
22 Ibid.
23 Ibid.
24 Overland and Poussenkova, *Russian Oil Companies in an Evolving World*, 5.
25 Ibid.
26 Bradshaw et al., "Preparing for the New Oil Order?"
27 Ibid.
28 Ibid.

29 Nina Tynkkynen, "A great ecological power in global climate policy? Framing climate change as a policy problem in Russian public discussion," *Environmental Politics*, 19(2), 179–95, 2010.
30 Alexandra Gillies, interview via phone, January 5, 2021.
31 Ibid.
32 Olúfẹ́mi O. Táíwò, interview via phone, January, 2021.
33 Benjamin Sovacool and Joseph Scarpaci, "Energy Justice and the Contested Petroleum Politics of Stranded Assets: Policy Insights from the Yasunî-ITT Initiative in Ecuador," *Energy Policy*, 95, 2016, 158–71.
34 Ibid.

9 Code

1 David Rolnick et al., *Tackling Climate Change with Machine Learning*, 2019, Cornell University, arxiv.org/pdf/1906.05433.pdf.
2 Restor, restor.eco, accessed April 15, 2021.
3 Rachel Frazin, "Government Probe Finds Companies Claiming Carbon Capture Tax Credit Didn't Follow EPA Requirements," *The Hill*, April 30, 2020.
4 Shoshana Zuboff, *The Age of Surveillance Capitalism: The Fight for a Human Future at the New Frontier of Power*, New York: PublicAffairs, 2019, 8 (italics in original).
5 Yuval Noah Harari, *Homo Deus: A Brief History of Tomorrow*, New York: Harper, 2017, 55.
6 Tim Hwang, *Subprime Attention Crisis: Advertising and the Time Bomb at the Heart of the Internet*, New York: Farrar, Straus and Giroux, 2020.
7 Benjamin Bratton, *The Terraforming*, London: Strelka Press, 2019, 10.
8 Ibid., 69, 63.
9 Jonathan Taplin, "Tech and Creative Destruction," in *Which Side of History? How Technology Is Reshaping Democracy and Our Lives,* ed. James P. Steyer. San Francisco: Chronicle Prism, 2020, 191.
10 Zephyr Teachout, *Break 'Em Up: Recovering Our Freedom from Big Ag, Big Tech, and Big Money*, New York: All Points Books, 2020.
11 Adrian Daub, *What Tech Calls Thinking: An Inquiry Into the*

Intellectual Bedrock of Silicon Valley, New York: FSG Originals, 2020.

12 Harari, *Homo Deus*, 15.

13 Microsoft, *Carbon Removal—Lessons from an Early Corporate Purchase*, 2021, 3. query.prod.cms.rt.microsoft.com/cms/api/am/binary/RE4MDlc.

14 Jaquelin Cochran and Paul Denholm, eds., *The Los Angeles 100% Renewable Energy Study*, Golden, CO: National Renewable Energy Laboratory, 2021.

15 Leigh Phillips and Michal Rozworski, *The People's Republic of Walmart: How the World's Biggest Corporations are Laying the Foundation for Socialism*. New York: Verso, 2019, 8.

16 Ibid., 244.

17 Daub, *What Tech Calls Thinking*, 46.

18 Ibid.,49.

19 Zero Cool, "Oil Is the New Data," *Logic*, no. 9, 2019.

20 Ibid.

21 Safiya Umoja Noble and Sarah T. Roberts, "Engine Failure: Safiya Umoja Noble and Sarah T. Roberts on the Problems of Platform Capitalism," *Logic*, no. 3: Justice, Winter 2017, 98.

22 Ben Tarnoff, "From Manchester to Barcelona," Logic, no. 9, Winter 2019, 100.

23 Teachout, *Break 'Em Up*.

24 Ibid.

25 Gabriel Winant, "No Going Back: The Power and Limits of the Anti-Monopolist Tradition," *The Nation*, January 21, 2020.

26 Tristan Harris, "Transforming the Attention Economy," in *Which Side of History: How Technology Is Reshaping Democracy and Our Lives*, ed. James P. Steyer, San Francisco: Chronicle Prism, 2020.

27 Ibid.

28 Philips and Rozworski, *The People's Republic of Walmart*, 7.

29 Hwang, *Subprime Attention Crisis*, 119.

30 Ibid., 120.

10 Political Power

1 Sem Oxenaar and Rick Bosman, "Managing the Decline of Fossil Fuels in a Fossil Fuel Intensive Economy: The Case of the Netherlands," in *The Palgrave Handbook of Managing Fossil Fuels and Energy Transitions*, edited by Geoffrey Wood and Keith Baker,

Cham, Switzerland: Springer Nature, 2020, 143.

2 Ibid.

3 Tobias Dan Nielsen, Karl Holmberg, and Johannes Stripple, "Need a Bag? A Review of Public Policies on Plastic Carrier Bags—Where, How and to What Effect?" *Waste Management*, 87, 2019.

4 Jennifer Clapp and Linda Swanston, "Doing Away with Plastic Shopping Bags: International Patterns of Norm Emergence in Policy Implementation," *Environmental Politics*, 18: 3, 2009, 315–32.

5 Ibid.

6 Bastian Loges and Anja P. Jakobi, "Not More Than the Sum of Its Parts: De-centered Norm Dynamics and the Governance of Plastics," *Environmental Politics*, 29: 6, 2020, 1004–23.

7 Jiajia Zheng and Sangwon Sun, "Strategies to Reduce the Global Carbon Footprint of Plastics," *Nature Climate Change*, 9, 2009, 374–8.

8 Jack Kaskey, "Oil Giants Bet Their Futures on Plastic—Just in Time for a Plastic-trash Crackdown," *Los Angeles Times*, June 9, 2019.

9 Nielsen et al., "Need a Bag?"

10 Cass Sunstein, "Montreal versus Kyoto: A Tale of Two Protocols," University of Chicago Law School, Public Law and Legal Theory Working Paper, no. 136, 2006.

11 United Nations Environment Program, "Kigali Amendment Implementation Begins," January 3, 2019, ozone.unep.org/kigali-amendment-implementation-begins.

12 United Nations Treaty Collection, "Status of Treaties," Chapter XXVII 2.f Amendment to the Montréal Protocol on Substances that Deplete the Ozone Layer, Kigali, 15 October 2016, accessed April 15, 2021.

13 Sunstein, "Montreal versus Kyoto."

14 Edward A. Parson, "The Technology Assessment Approach to Climate Change," *Issues in Science and Technology*, 18: 4, 2002, 65–72.

15 Davis Cyranoski, "China Feels the Heat over Rogue CFC Emissions," *Nature*, 571: 7765, 2019, 309.

16 Charles Herrick, "A Stratospheric Success," *Nature*, 422, 2003, 664–65.

17 Ruth Malone, "The Race to a Tobacco Endgame," *Tobacco Control*, 25: 6, 2016.

18 Elizabeth A. Smith, "Questions for a Tobacco-free Future," *Tobacco Control*, 22, 2013.
19 Malone, "The Race to a Tobacco Endgame."
20 Smith, 2013, 1.
21 Ruth Roemer, Allyn Taylor, and Jean Larviere, "Origins of the WHO Framework Convention on Tobacco Control," *American Journal on Public Health* 95: 6, 2005, 936–8.
22 Ibid.
23 Ivetta Gerasimchuk and James Bacchus, "To Phase out Coal, World Leaders Should Learn from Tobacco Action," IISD blog, December 11, 2017.
24 Doug Kysar, "Fossil Fuel Industry's 'Tobacco Moment' Has Arrived," *Law360*, 2017.
25 Eric Larson et al., *Net-Zero America: Potential Pathways, Infrastructure, and Impacts*, interim report, Princeton, NJ: Princeton University, December 15, 2020.
26 **Xiaowei** Wang, *Blockchain Chicken Farm and Other Stories of Tech in China's Countryside,* New York: Farrar, Straus and Giroux, 2020, 2.

11 Moratoria, Bans, and Refusal to Finance

1 Andreas Malm, *How to Blow Up a Pipeline*, New York: Verso, 2021, 8.
2 Ibid., 67.
3 Nick Estes, *Our History Is the Future: Standing Rock Versus the Dakota Access Pipeline, and the Long Tradition of Indigenous Resistance,* London: Verso Books, 2019; Dina Gilio-Whitaker, *As Long as Grass Grows: The Indigenous Fight for Environmental Justice from Colonization to Standing Rock,* Boston: Beacon Press, 2019.
4 The Red Nation, "TRN Statement on US Elections 2020," November 15, 2020, therednation.org/trn-statement-on-us-elections-2020.
5 Nathan Ratledge, Steven J. Davis, and Laura Zachary, "Public Lands Fly under Climate Radar," *Nature Climate Change*, 9: 2, 2019, 92–3.
6 Peter Erickson, Michael Lazarus, and Georgia Piggot, "Limiting Fossil Fuel Production as the Next Big Step in Climate Policy," *Nature Climate Change*, 8: 12, 2018, 1037–43.
7 Nicolas Gaugin and Philippe Le Billion, "Climate Change and Fossil Fuel Production Cuts: Assessing Global Supply-side

Constraints and Policy Implications," *Climate Policy*, 20: 8, 2020, 888–901.

8 Daniel Yergin, *The New Map: Energy, Climate and the Clash of Nations,* New York: Penguin, 2020, 55.

9 Tim Donaghy, *Policy Briefing: Carbon Impacts of Reinstating the U.S. Crude Export Ban*, Greenpeace USA Policy Briefing, January 2020, greenpeace.org.

10 John Noël, interview, November 2020.

11 Jonas Meckling and Jonas Nahm, "The Politics of Technology Bans: Industrial Policy Competition Goals for the Auto Industry," *Energy Policy*, 126, 2019, 470–9.

12 Associated Press, "Restaurant Group Sues over Berkeley's Natural Gas Ban," November 21, 2019.

13 Cleo Verkuijl, N. Jones, and M. Lazarus, *Untapped Ambition: Addressing Fossil Fuel Production through NDCs and LEDS* (SEI Working Paper), Stockholm: Stockholm Environment Institute, 2019.

14 Georgia Piggot et al., "Swimming Upstream: Addressing Fossil Fuel Supply under the UNFCCC," *Climate Policy*, 18: 9, 2018.

15 Peter Newell and Andrew Simms, "Towards a Supply Side Climate Treaty," *Climate Policy*, 20: 8, 2020, 1043–54.

16 Kyla Tienhaara, interview, December 2020.

17 Kyla Tienhaara and Lorenzo Cotula, *Raising the Cost of Climate Action? Investor–state Dispute Settlement and Compensation for Stranded Fossil Fuel Assets.* IIED Land, Investment and Rights Series, 2020, pubs.iied.org.

18 Transnational Institute, "Busting the Myths around the Energy Charter Treaty," December 15, 2020, tni.org/en/ect-mythbuster.

12 Ending Subsidies

1 Tim Rayner, "Keeping it in the ground? Assessing global governance for fossil-fuel supply reduction," *Earth System Governance*, online first, October 1, 2020.

2 Ivetta Gerasimchuk et al., *A Guidebook to Reviews of Fossil Fuel Subsidies: From Self-reports to Peer Learning,* Winnipeg: International Institute for Sustainable Development, 2017.

3 Peter Erickson et al., "Why Fossil Fuel Producer Subsidies Matter," *Nature*, 578: 7793, 2020, E1–4.

4 Ibid.

5 Jessica Jewell et al., "Limited Emission Reductions from Fuel

Subsidy Removal Except in Energy-exporting Regions," *Nature*, 554: 7691, 2018, 229–33.

6 Ivetta Gerasimchuck et al., *Zombie Energy: Climate Benefits of Ending Subsidies to Fossil Fuel Production*, International Institute for Sustainable Development working paper, 2017.

7 Jewell et al., "Limited Emission Reductions."

8 D. Coady et al., "How Large Are Global Fossil Fuel Subsidies?" *World Development*, 91, 2017, 11–27.

9 Irene Monasterolo and Marco Raberto, "The Impact of Phasing out Fossil Fuel Subsidies on the Low-carbon Transition," *Energy Policy* 124, 2019.

10 Francisco Javier Arze del Granado, David Coady, and Robert Gillingham, "The Unequal Benefits of Fuel Subsidies: A Review of Evidence for Developing Countries," IMF Working Paper, 2010.

11 IISD blog, "How Reforming Fossil Fuel Subsidies Can Go Wrong: A Lesson from Ecuador," October 24, 2019.

12 P. Gass and D. Echeverria, *Fossil Fuel Subsidy Reform and the Just Transition*, Winnipeg: International Institute for Sustainable Development, 2017.

13 Permission to Extract

1 Ethan Elkind and Ted Lamm, *Legal Grounds: Law and Policy Options to Facilitate a Phase-Out of Fossil Fuel Production in California*, Berkeley: Center for Law, Energy and the Environment, 2020.

2 Thijs Van de Graaf and Aviel Verbruggen, "The Oil Endgame: Strategies of Oil Exporters in a Carbon-Constrained World," *Environmental Science & Policy*, 54, 2015, 456–62.

3 European Commission, "The End of Sugar Quotas in the EU," 2017, ec.europa.eu/commission/presscorner/detail/en/MEMO_17_3488.

4 Emmet Livingstone, "Europe Offers 500 Million to Help Dairy Farmers," *Politico*, July 15, 2016.

5 Richard Durbin and John Kennedy, Letter to DEA Acting Administrator Timothy Shea, July 30, 2020, durbin.senate.gov/newsroom/press-releases/durbin-kennedy-urge-dea-to-use-authority-to-reduce-excessive-opioid-production-quotas.

6 Ibid.

7 Michael Schatman and Erica Wegrzyn, "The United State Drug

Enforcement Administration and Prescription Opioid Production Quotas: An End Game of Eradication?" *Journal of Pain Research*, 13, 2020, 2629–31.

8 Geir B. Asheim et al., "The Case for a Supply-side Climate Treaty," *Science* 365: 6451, 2019, 325–7.

9 Andreas Malm, "Planning the Planet: Geoengineering Our Way Out of and Back into a Planned Economy," in *Has It Come to This? The Promise and Peril of Geoengineering on the Brink*, eds. J. P. Sapinski, Holly Buck, and Andreas Malm, New Brunswick, NJ: Rutgers University Press, 2020, 144.

10 Ibid, 146.

11 Laurence Delina, *Strategies for Rapid Climate Mitigation: Wartime Mobilisation as a Model for Action*, New York: Routledge, 2020.

12 Andreas Malm, *Corona, Climate, Chronic Emergency*, New York: Verso Books, 2020, 155.

13 Ibid., 163 (italics in original).

14 Nationalize for Exit

1 Gar Alperovitz, Joe Guinan, and Thomas M. Hanna, "The Policy Weapon Climate Activists Need," *The Nation*, April 26, 2017.

2 Alexander Kaufman, "Falling Oil Prices Breathe New Life into an Old Idea: Nationalize the Industry," *Grist*, April 20, 2020.

3 David S. Painter, "Oil, Resources, and the Cold War, 1945–1962," in *The Cambridge History of the Cold War,* edited by Melvyn P. Leffler and Odd Arne Westad. New York: Cambridge University Press, 2010.

4 Ibid.

5 Ekim Arbatli, "Resource Nationalism Revisited: A New Conceptualization in Light of Changing Actors and Strategies in the Oil Industry," *Energy Research and Social Science* 40, 2018, 101–108.

6 Ibid.

7 Thomas Hanna, "A History of Nationalization in the United States: 1917–2009," The Next System Project, 2019.

8 Peter Gowan, "A Plan to Nationalize Fossil Fuel Companies," People's Policy Project, March 21, 2018.

9 Daniel Yergin, *The New Map: Energy, Climate and the Clash of Nations,* New York: Penguin, 2020.

10 Rachel Maddow, *Blowout: Corrupted Democracy, Rogue State*

Russia, and the Richest, Most Destructive Industry on Earth, New York: Crown, 2019, 39, 41.

11 Alexandra Gillies, interview via phone, January 5, 2020.

12 Marcela Mulholland and Ethan Winter, "Nationalize the Fossil Fuel Industry," Data For Progress, April 21, 2020.

13 Sean Sweeney, "There May Be no Choice but to Nationalize Oil and Gas—and Renewables, Too," *New Labor Forum*, 29(3), August 2020, 114–20.

14 Ibid.

15 Reverse Engineer

1 Rhodium Group, *Capturing New Jobs: The Employment Opportunities Associated with Scale-up of Direct Air Capture (DAC) Technology*, 2020, rhg.com/research/capturing-new-jobs-and-new-business.

2 Rachel Cohen, "The Environmental Left Is Softening on Carbon Capture," *The Intercept*, September 20, 2019.

3 BlueGreen Alliance, *Solidarity for Climate Action*, 2019, bluegreenalliance.org.

4 Myles R. Allen, David J. Frame, and Charles F. Mason, "The Case for Mandatory Sequestration," *Nature Geoscience*, 2: 12, 2009, 813–14.

5 Olúfẹ́mi O. Táíwò, interview via phone, January, 2021.